Ore Textures

Roger Taylor

Ore Textures

Recognition and Interpretation

Springer

EGRU

Roger Taylor
Townsville
Queensland 4810
Australia

Additional material to this book can be downloaded from http://extras.springer.com.

ISBN 978-3-662-50213-6 ISBN 978-3-642-01783-4 (eBook)
DOI 10.1007/978-3-642-01783-4
Springer Dordrecht Heidelberg London New York

Cover design: Bauer, Thomas

Printed on acid-free paper

Springer is part of Springer Science+Business Media (www.springer.com)

Acknowledgments

Given the eleven years required to produce the original five volumes, the list of contributors is considerable. They constitute numerous individuals and financial sponsor companies who deserve special acknowledgement.

It is impractical to give lengthy written credit to all concerned, and the list below only outlines some of their contributions. Thank you everybody.

Special thanks are extended to Dr Peter Pollard and Dr Gavin Clarke who contributed continuously via encouragement, editing, rock preparation, and their own extensive knowledge of ore textures.

The outstanding plates are the work of Roger Yeldham and Brian Pump from the James Cook University photographic division. Although not obvious here, David Lloyd contributed the initial graphic design which made the individual volumes so attractive. The difficult and delicate task of combining five volumes into a single offering has been conducted by Elisabeth Sillmann of blatterwald Design. Her high quality graphic design skills are gratefully acknowleged, and will be evident to all readers. Some anonymous frogs who gracefully permitted use of their photographic images in the original volumes are also acknowledged (unfortunately only two could be included here).

INDIVIDUAL CONTRIBUTORS

Dr A. Allibone	Specimen loan and stuctural expertise
Mr D. Andrews	Specimen loan and photography
Dr E. M. Baker	Photography
Professor T. H. Bell	Structural advice
Dr N. Adshead-Bell	Structural advice
Dr S. Beams	Specimen loan
Ms E. Caragianides	Sundowners
Ms D. Casey	Word processing and fund raising
Dr G. W. Clarke	Specimen loan, specimen preparation, photography and textural expertise
Dr G. Dong	Specimen loan
Ms M. Drögemuller	Word processing and fund raising
Dr K. Hickey	Structural advice
Mrs K. Hotschilt	Photography
Dr W. P. Laing	Specimen loan
Mrs L. Laivens	Word processing
Ms H. Lawrie	Editing
Mr D. Lloyd	Graphic design
Ms F. McKinstry	Frog photograph
Ms M. Meyers	Editing
Dr R. Meyers	Editing and general advice
Dr G. W. Morrison	Specimen loan, photography and textural expertise
Dr B. New	Structural advice
Professor G. N. Phillips	Specimen loan and general advice
Dr P. J. Pollard	Specimen loan, editing, photography, textural expertise and general advice
Mr B. Pump	Photography
Mr S. Richards	Frog photograph
Ms E. Sillmann	Graphic design, typesetting
Dr P. J. Williams	Specimen loan and general advice
Associate Professor S. E. Williams	Frog photograph
Dr P. J. Wormald	Specimen loan
Dr M. Worsley	Specimen loan
Mr R. Yeldham	Photography

COMMERCIAL SPONSORS

Aberfoyle Resources Limited

ABX Exploranciones

Acacia Resources Limited

Anglo American Exploration (Australia) Pty Ltd

ARIMCO Mining Pty Ltd

Aurora Gold Pty Ltd

Australasian Institute of Mining and Metallurgy (North Queensland Branch)

Aztec Mining Company Ltd

Battle Mountain (Australia) Inc

Barrick Gold Corporation (Australia)

BHP Pty Ltd

BHP Billiton (Cannington)

Billiton Australia

CRA Exploration Pty Ltd

Companhia Vale Do Rio Doce (CVRD)

Cyprus Gold Australia Corp.

Delta Gold NL

Department of Minerals & Energy (Queensland)

Department of Minerals & Energy, PNG

Department of Resource Industries

Dominion Mining Ltd

Economic Geology Research Unit (EGRU) James Cook University

Freeport McMoran Copper and Gold Co

Giants Reef Mining

Geopeko

Goldfields Exploration

Gold Mines Of Australia

Great Central Mines and Centaur Mining and Exploration Ltd

Homestake Australia Ltd

Ivanhoe Mines Ltd

James Cook University

Metana Minerals NL

MIM Exploration Pty Ltd

Mount Isa Mines Ltd

Mount Leyshon Gold Mines Ltd

Newcrest Mining Ltd

Newmont Gold Company

Normandy Exploration Ltd

North Ltd

North Flinders Exploration

Nuigini Mining (Australia) Pty Ltd

Pac Min Mining Company Ltd

Pancontinental Mining Ltd

Pasminco Exploration

Placer Exploration Ltd

Plutonic Mining

Phelps Dodge Australia Inc

P. T. Freeport Indonesia

Reynolds Australia Metals Ltd

RGC Ltd

Rio Tinto

Ross Mining NL

RTZ (South America)

Sons of Gwalia Ltd

Teck Cominco Ltd

The Key Centre for Economic Geology, James Cook University

Western Mining Corporation

Preface

This text has resulted from some forty years of experience during which the author has puzzled over the meaning of ore textures. The learning process has been slow and is still incomplete. The bemusement began directly upon leaving the academic confines, which in retrospect left one keen young geologist very ill-equipped to interpret the mineralising process via field or hand lens style observation of the rocks. Enlightenment has proceeded via a series of events:-

1. The slow process of field observation as an industry-based mine and exploration geologist.
2. The opportunity to visit numerous different styles of ore deposit both as an academic and consultant.
3. The need to answer questions from several generations of enquiring students.
4. The privilege of being able to conduct research.
5. The good fortune to be associated with a few top class economic geologists who actually knew what they were doing!

Professor Willard C. Lacy deserves the most credit for quietly demonstrating the value of first principles concerning fluid channelways, combined with an ability to look properly at the textures. Despite rapid technological advances this skill remains fundamental both to the practicing exploration geologist and to the academic researcher who is interested in understanding ore forming processes.

The five volumes comprising this presentation were originally compiled individually over a period of eleven years, and aimed to progressively guide the observer though the principles of recognition concerning infill, alteration, overprinting, and finally to ore-related breccias.

Most observers have no significant problems with infill and alteration concepts in relation to narrow veins, but frequently flounder when the same components are repackaged as breccias.

The plates have been carefully chosen to illustrate recognition procedure whilst giving a visual guide to a wide range of examples. The plate captions have been specifically written in a lengthy simplistic form to assist the genuine beginner. They contain a wealth of detail which is intended to lay a foundation for good textural observational methodology. A few more complex plates have been sporadically included to add appeal for those at connoisseur level.

The book is not intended as a rigorous academic text. However, it is noted that the subject of textural interpretation still remains absent from most institutional training courses, and it is thus hoped that this contribution will be of direct assistance to many enquiring readers.

Townsville, March 2009 *Roger G. Taylor*

Contents

Ore Textures – Recognition and Interpretation

Volume 3 • Overprinting Textures 99

Volume 1 – Infill Textures

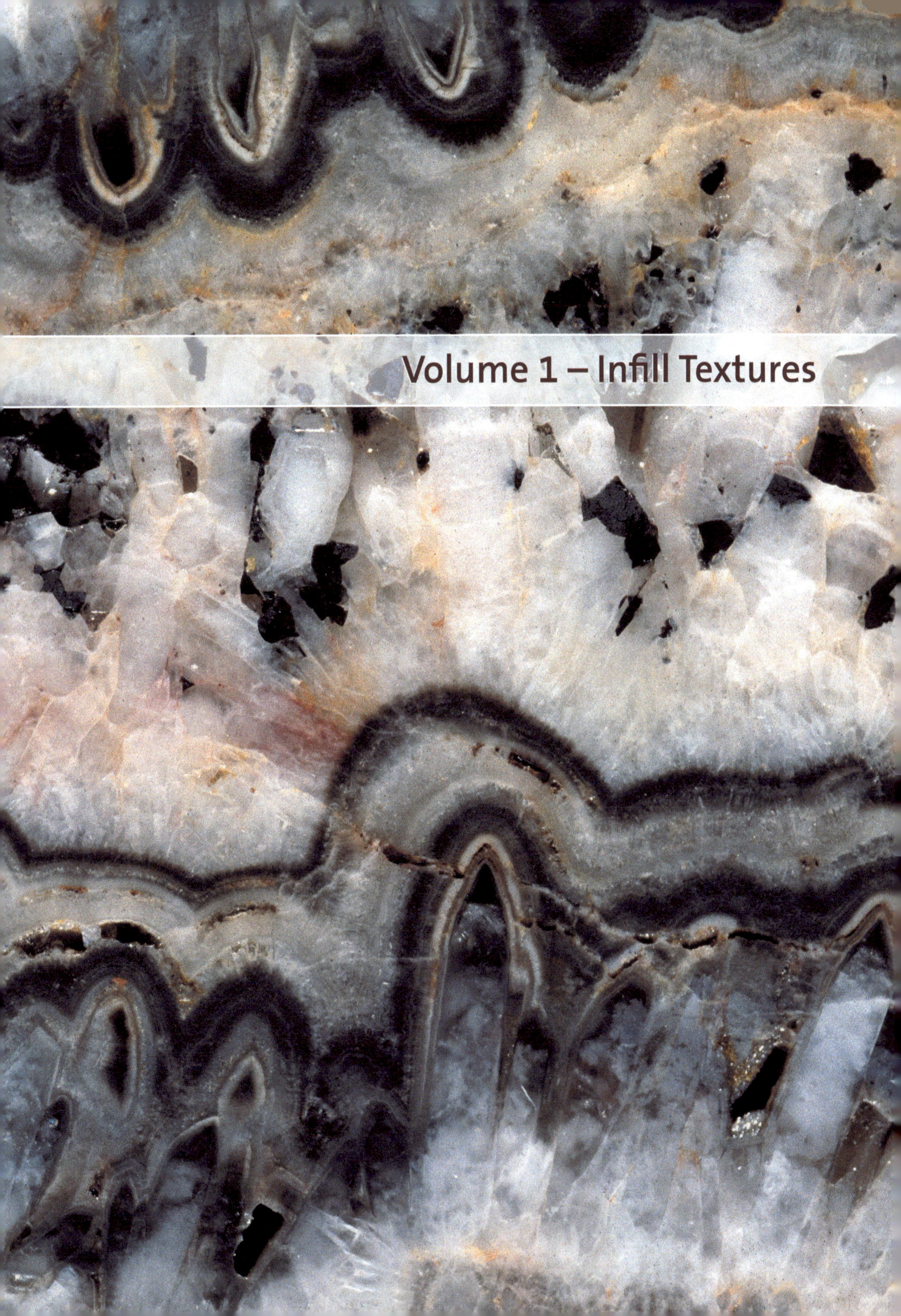

Volume Front Page

Specimen of vein material from the St Patrick tin mine, near Irvinebank, Queensland (Herberton tinfield), Australia. The specimen illustrates many of the criteria which can be utilised to identify infill mineralogy:

(a) Cavities (several styles)
(b) Cavities containing partially euhedral minerals (quartz)
(c) Euhedral (or partially euhedral) crystals (quartz and dark cassiterite)
(d) Layered textures – crustiform banding (quartz)
(e) Triangular textures – small scale acute angle forms of cassiterite (dark) towards the top centre where cassiterite has infilled between pre-existing quartz crystals.

The wall rocks are silicified metasediments (not visible). The paragenesis (depositional sequence) is obviously complex, with the many different forms of silica presumably reflecting changing physical and chemical conditions of precipitation. It is possible that the thin curving style cavities which are prevalent towards the bottom of the plate have formed by selective leaching of an infill mineral (siderite?).

1 Introduction

The recognition and interpretation of ore textures is a difficult task, and the problem is compounded firstly by the relatively low priority given to the subject within most tertiary training institutions, and secondly by the lack of simple well illustrated texts. Special reference needs to be focused upon features which are visible to the eye or observed via a hand lens. These are the most valuable to practicing field geologists.

Most epigenetic ore deposits result from hydrothermal fluids which are chanelled through zones of permeability, with the various ore-related components being formed either by precipitation in void space, or by interaction with the wall rocks. The permeability controls (channelways) vary widely in detail ranging from primary permeability in unconsolidated sediments through to secondary permeability imposed by various styles of fracturing. In all instances the system will contain components derived from precipitation within the fluid-filled void spaces (*Infill*) and components derived by reaction with the wall rocks (*Alteration*). Commonly the combination masks the nature of the original channelways and from this perspective it is *crucial to separate the two components to comprehend the nature of the permeable zone.* The separation of the components also leads to a comprehension of the nature of the fluids involved. It could be fairly stated that geologists wishing to comprehend ore deposits should "chase the channelways" by carefully distinguishing between infill and alteration components.

This contribution is aimed at the question of how to recognise infill textures. Such a proposition may sound a little simplistic but infill textures present in a wide variety of different disguises, and frequently elude the inexperienced observer. A considerable amount of information is contained within the plate descriptions which have been deliberately written at some length to guide the reader in both the process and practice of observation. It is envisaged that this volume will be the first of a series on the subject of textural recognition.

2 Recognition Criteria

2.1 Incomplete Infill

2.1.1 Remaining Void Space

This is the most obvious criterion focusing the attention of the observer and signaling the probable location of a permeability channelway.

In most cases the precipitating minerals completely fill the space traversed by the ore fluids. However there are numerous circumstances where the process is incomplete and the presence of any form of cavity within a mineralised system would immediately come under suspicion as marking a region of infill. Cavity recognition is simple enough at the macroscale but frequently overlooked at the hand lens scale. A striking example is illustrated by ***Plate 1*** and other examples are present in ***Plates 8, 9, 21 and 23.***

2.1.2 Crystals Projecting Into a Cavity

Many cases examples of remaining void space also contain well-formed crystals projecting into the cavity. Quartz crystals are particularly common, as are typically late phase low temperature minerals such as calcite, siderite, and fluorite. Pyrite and marcasite are well-represented from the sulphide group.

Again this criterion is easily recognised at the coarse scale but is commonly missed when the cavities are small. A good example is shown on ***Plate 1,*** whilst less obvious situations are present on ***Plates 8, 9, 21 and 23.***

◻ PLATE 1

VOIDS, PROJECTING CRYSTALS, EUHEDRAL CRYSTALS, Edge problems, working backwards.

This specimen illustrates the very obvious criterion of a cavity which represents an unfilled portion of the original fluid filled space. It also contains well-depeloped quartz crystals projecting into the cavity.

The specimen also illustrates a good working principle which is discussed later in this manual i.e. the technique of working backwards from the known. The quartz crystals need a substrate to nucleate upon and careful inspection will reveal that they nucleate off a pale-buff, siliceous base. This is also infill. A further principle is well-illustrated here concerning edge problems and breccia recognition. The cavity is actually one of a series of small holes in the rock, which is a breccia. The fragments have been extensively silicified and can just be discerned on the plate as the darker, mottled zones in the top left and bottom right hand corners. The junction between silicified fragments and the buff-coloured, silica infill is naturally a little vague.

Note:

It is always prudent to establish the origin of the cavities with respect to weathering. Late-phase carbonates or sulphides which are common, final stage infill products are easily leached during surficial oxidation. In most cases their existence is fingerprinted by the presence of limonite gossan or iron staining in, and around, the cavity. This natural etching usually enhances infill recognition and produces cavities with projecting crystals. The clean nature of the cavities in the above plate suggests that they never completely filled.

2.2 Complete Infill

The absence of voids or free crystals makes the task of recognising infill a little more difficult. However in the majority of cases the following criteria will prove of value. It should be noted that some of the following criteria are not necessarily diagnostic, but utilised in conjunction with each other and tempered with geological common sense, they can be regarded as good indicators. Some of the potential problems are outlined later.

2.2.1 Euhedral or Partially Euhedral Crystal Outlines

Since the fluid filled void offers an excellent opportunity for unhindered crystal growth it is very common for cavity infill minerals to achieve good crystal forms. This is especially true of quartz which is probably the most common mineral in hydrothermal systems. Quartz very rarely adopts its familiar hexagonal form other than by crystallisation in open space and hence any hint of a quartz crystal is to be regarded as a potential indicator of a fluid channelway. Similar comments would apply to many other common hydrothermal minerals such as feldspars, fluorite, cassiterite, galena, sphalerite and chalcopyrite. Some care is required with those minerals which are very capable of forming euhedral shapes via alteration/replacement such as pyrite, arsenopyrite and carbonates. However, in general well-formed crystals are excellent signposts for infill recognition. Numerous examples are scattered through the illustrations with good examples shown on *Plates 2, 3, 4, 5 and 6*. Less obvious examples are shown on *Plates 17, 18 and 22*.

◻ PLATE 2

EUHEDRAL CRYSTALS, triangular textures, small scale, irregular nucleation, irregular shapes in cross section.

The specimen is included to demonstrate euhedral mineral growth within a vein-style open space and is composed totally of infill.

The most striking components are the large quartz crystals, one of which exhibits clear growth zoning. They have grown from the substrate at different angles and hence present different shaped cross sections when slabbed. They are irregularly overgrown by subhedral to euhedral dark cassiterite crystals which illustrate the point that not all nucleating products form nice layers. Erratic nucleation is quite common where isolated crystals nucleate at seemingly random sites. Silica deposition has continued at some point in time to completely fill the original cavity. The pale yellowish orange mineral is probably siderite, most of which is clearly late and forms very good triangular-acute angle textures as it infills between quartz crystals. The specimen also demonstrates a well-defined paragenetic sequence of deposition, and recognition of infill textures is vital in any paragenetic

PLATE 1 The remaining void space and crystals projecting into a cavity criteria.
Specimen from Mt Gibson (tin, topaz) region, near Mt Gamet, Queensland, Australia.

PLATE 2 The euhedral mineral criterion
Specimen from Mt Misery – Mt Tin – Morning Cloud deposit, Mowbray Creek mineralisation centre, near Herberton tinfield, Queensland, Australia.

study to establish timing. Surprisingly large numbers of paragenetic studies not only fail to distinguish infill from alteration, but seem unaware that without such distinction the results are of limited value. It should be noted that the sequence of infill deposition only establishes relative timing. The carbonate in the above picture may have deposited a very long time after the other minerals and might even belong to a different paragenetic stage.

◘ PLATE 3

EUHEDRAL CRYSTALS, zoning, large scale triangular texture.

The wolframite (black) crystals have precipitated very early in the paragenetic sequence and have suffered minor deformation and replacement from later processes. Potassium feldspar (pink) has infilled interstitially around some of the crystals which have been fractured either before or at the time of feldspar deposition. It is possible that minor alteration of the wolframite to feldspar has occurred within the fractured zones as there are many cases where the walls on either side of the fracture do not match. Alternatively the wolframite crystal may have been affected by dissolution prior to feldspar precipitation. Massive quartz surrounds the wolframite and despite the lack of texture would be assumed to represent infill. This suspicion is supported by the presence of barely discernible, finely layered growth zones which are visible above the silver-coloured (cleavage reflectance) wolframite crystal towards the top left of the plate.

Late fracturing has lead to the introduction of sulphides which are associated with a subtle grey alteration (silica?). This is well-developed in the feldspar crystals. The alteration halo together with the triangular-acute angle texture of the sulphide patches is strongly suggestive of infill, although there could be minor sulphide alteration/replacement within the fractured wolframite crystals.

◘ PLATE 4

EUHEDRAL CRYSTALS, small scale triangular texture.

The yellow chalcopyrite also exhibits numerous examples of the small-scale triangular-acute angle texture resulting from infilling around/between pre-existing euhedral minerals. Several unidentified minerals are also present including a resinous euhedral mineral and a late? dark interstitial phase. Although not shown in this plate the specimen is from a breccia zone where anastomosing open-space infill of coarse-grained magnetite and chalcopyrite surrounds altered fragments of a fine-grained, magnetite-rich host rock (ironstone).

The ironstone forms part of a pre-existing stratigraphy which has been extensively folded and subjected to considerable deformation/metamorphism. The infill depicted in *Plate 4* is undeformed.

2.2.2 Zoned Crystals

Finely zoned crystals are particularly difficult to develop via alteration/replacement. Hence crystals or hints of crystals with well developed growth zoning are particularly useful indicators of infill. Unfortunately such growth zoning is usually difficult to see in most minerals at the hand specimen level. Fortunately the phenomenon is commonly well-developed in quartz, and all suspected hexagonal cross sectional examples merit careful inspection. Examples are illustrated in *Plates 2 and 3*.

It should be noted that growth zoning via seqential diffusion is common in some skarn minerals growing via alteration of carbonate (especially garnet).

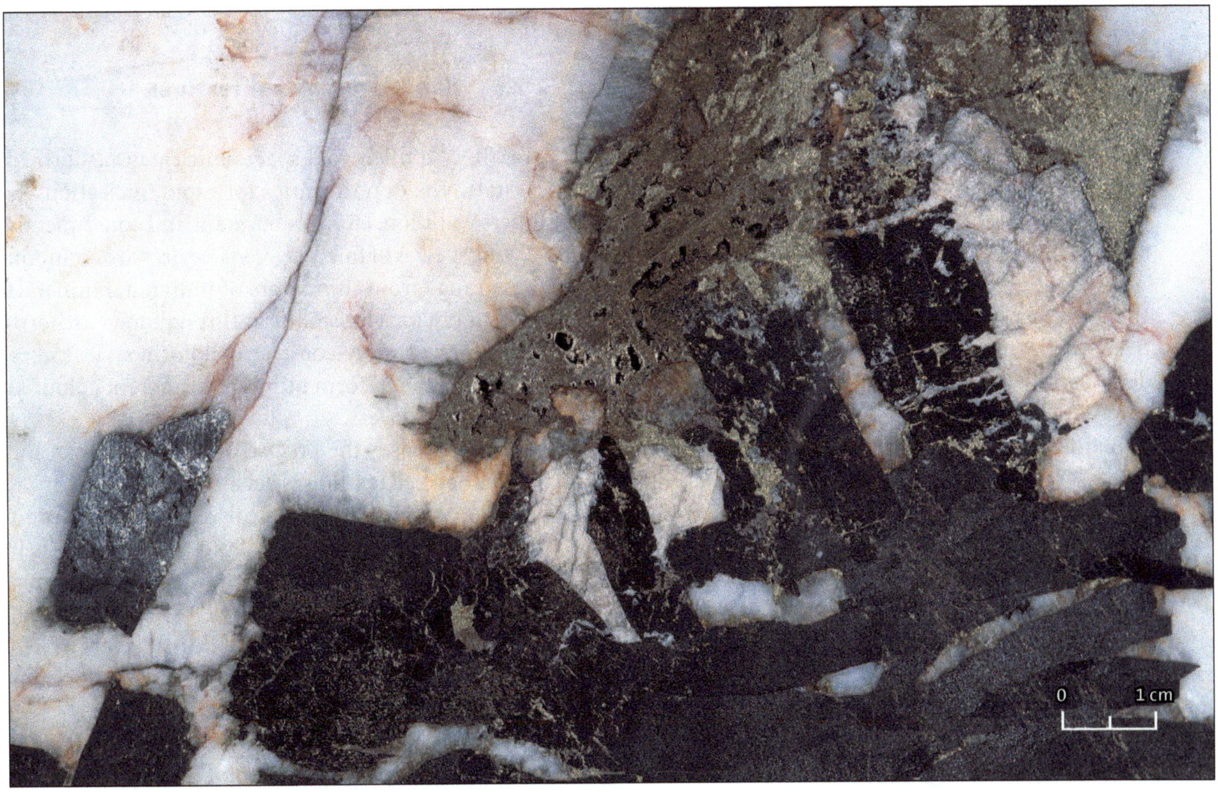

PLATE 3 The euhedral mineral criterion.
The location of the specimen is uncertain but probably from Mt Carbine tungsten mine, Mt Carbine, Queensland, Australia. The plate is presented to show euhedral crystals of wolframite within a quartz vein.

The euhedral mineral criterion.
PLATE 4 Specimen from Selwyn-Cloncurry region, Queensland, Australia.

PLATE 5

MAIN CRITERIA/TECHNIQUES ILLUSTRATED: EUHEDRAL CRYSTALS, small scale triangular textures, working backwards (and forwards), irregular nucleation.

The specimen has been selected to illustrate a relatively subtle euhedral growth of quartz and to further demonstrate the inferences that can be drawn from recognition that it represents infill.

A line of euhedral to subhedral quartz crystals (white-grey) and dark crystals?, can be seen down the centre-right of the plate. Given that quartz very rarely forms euhedral shapes via alteration/replacement processes it is reasonable to assume that it represents infill nucleating from the brown material beneath (cassiterite)·

It follows that the – adjacent – sulphide components of chalcopyrite (yellow) and pyrrhotite (bronze) are probably also infill, and represent a vein style channelway.

Careful observation of the region immediately beneath to the left of the quartz crystals reveals an intermittent, small, dark-grey shadow effect which is probably siliceous (or possibly amphibole) alteration of the cassiterite.

This same alteration effect can be seen in several places towards the left of the brown layer where it is again associated with dark crystals projecting into areas of chalcopyrite + pyrrhotite exhibiting good triangular-acute angle textures. All this suggests that earlier cassiterite has been brecciated, with the resulting open spaces infilled by silica-chalcopyrite-pyrrhotite. Alternatively, the cassiterite may also represent infill material which contains residual, open space subsequently occupied by the later minerals.

The left region of the plate is obviously more complex and difficult to interpret. It is possible that the dark siliceous zones are highly altered fragments of wall rocks. However this would require confirmation via further sample acquisition.

It is not suggested that the above interpretation is totally correct, but there is ample evidence to suspect a substantial amount of infill, leading from the recognition of the quartz crystals. Obviously a serious research study would acquire more specimens and seek confirmation via both this process and microscopy.

2.2.3 Banded or Layered Textures

Not all fluid filled void spaces nucleate well-formed crystals. The opportunities for rapid nucleation via supersaturation, etc. are abundant and consequently various species of finely layered textures are common. These range from thin layers of individual minerals without obvious crystal form through to crustiform bands or even more complex colloform styles. Silica in particular can form numerous different coloured and textured layers.

In many cases the precise origin of many of the banded/layered textures is debatable, but for practical purposes they are exceptionally common end-products of crystallisation within fluid-filled space. Their propensity to occur within shallow level (epithermal and surficial) environments is well known although they are also well-represented at deeper levels. When they form around pre-existing fragments they commonly produce eye-catching results.

Some of the finely layered textures can also be produced via replacement (concretions, pisolites etc), but utilising geological common sense it is usually possible to identify these. Problems also exist with various forms of direct precipitation of the evaporitic, banded ironstone, or volcanogenic styles. However, when put into geological context there are relatively few situations where the observer experiences significant problems. It would be fair to state that any form of layering or banding should be seriously considered as a potential infill texture. Good examples of infill layering are depicted in *Plates 6, 7 and 8*.

PLATE 6

EUHEDRAL CRYSTALS, layering.

This complex specimen has been selected to show a less-obvious euhedral crystal form combined with layering.

The crystal form here is represented by the early formed dark, equant sphalerite which has been overprinted by at least two successive crustiform coatings of later sphalerite (pale yellow and brownish yellow). Late quartz (white) has infilled around the sphalerite clusters and there is little doubt that the entire specimen represents an infilled open space.

The above features are subtle and illustrate the need for careful inspection. But we can see them in the picture which is not hand lens scale!

PLATE 5 The euhedral mineral criterion.
Specimen of vein? material from Cleveland tin mine, Tasmania, Australia.

Well-formed crystals and layered textures as infill criteria.
Specimen of vein material from the Snow Flake mine, San Juan (silver, lead, zinc, gold) district, Colorado, United States of America.
(Specimen provided by G. W. Morrison).

PLATE 6

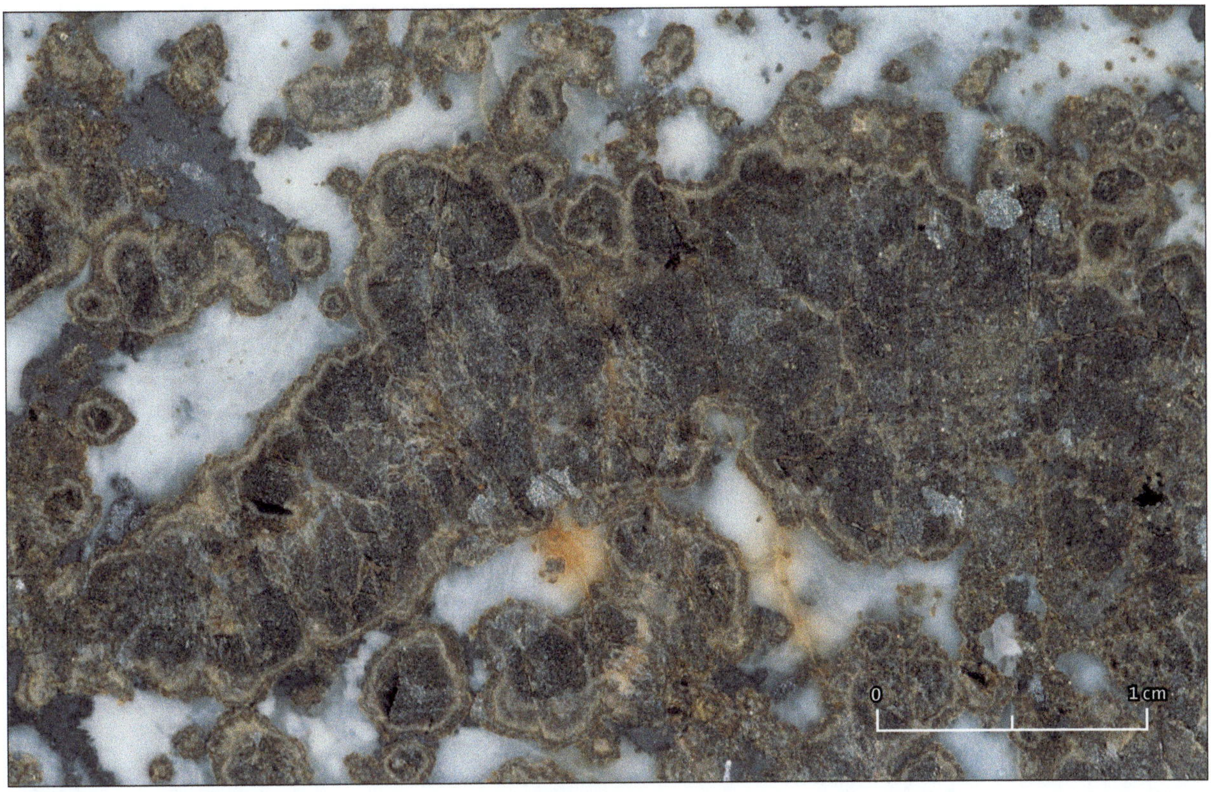

It is equally clear that the specimen exhibits some complexities that would require resolution via thin section examination. The boundaries between the pale-yellow and dark, crystalline sphalerite are commonly rather vague, suggesting that some form of alteration from dark to pale-yellow may have also occurred. The status of the silver coloured minerals (galena? sulphosalts?) is also difficult to decipher. Some may pre-date the crustiform sphalerite whilst the majority are clearly later and form post-sphalerite but pre-quartz.

It is the function of this manual to present information that can be seen via the eye or hand lens, and some of the finer paragenetic points require either more specimens or microscopic study. However it is clear that the paragenesis is resolvable and that the textures have resulted from the infilling of a fluid-filled clannelway over an unspecified length of time.

◘ PLATE 7

LAYERING, euhedral crystals.

This attractive specimen, although complex paragenetically has been included to show a variety of crustiform styles of silica.

Numerous generations of silica are represented within the plate and this complex crustiform layering is very common in high-level epithermal environments. It can however, also form at deeper crustal levels. It is usually interpreted as representing repeated, rapid nucleation due to extremely changeable conditions such an intermittent boiling/pressure change, etc.

Fragments of pre-existing silica rocks are coated with layers of crustiform, microcrystalline silica to produce a species of cockade texture. The fragments also contain varieties of crustiform and more coarsely crystalline quartz.

Some of the late-stage infill is more coarsely crystalline and there are many examples of triangular-acute angle textures of dark material filling in between the larger crystals. A good example is present towards the top left where small, white, bladed crystals interlock to resemble scattered matchsticks. The crystals are composed of white silica but are thought to have been bladed carbonate which has been silicified. The coarser, pale buff-grey material (top, centre-right) is also considered to represent a large cluster of silicified, bladed carbonate. The "carbonate" texture is a common feature of epithermal veins.

2.2.4 Large-Scale Triangular Textures

Large-scale triangular textures are particularly common in brecciated rocks and represent void spaces between rock fragments. These may be partially infilled in which case they are relatively easy to recognise, or completely infilled, which renders their observation a little more difficult. In many cases the passage of hydrothermal fluids also completely alters the host fragments which presents a major barrier to field recognition. This problem is further exacerbated when the alteration minerals are very similar to the infilling minerals (e. g. silica infill accompanied by silica alteration).

The term "triangular" needs some explanation. Obviously, cavities formed via breaking rocks will take on a variety of shapes depending on the angularity of the rocks, the amount of rock matrix, the range in size of rock fragments etc. The term "acute angled" may be more appropriate for the cross sectional representation of such cavities, although even this can become a little inappropriate for some situations. The resulting shapes are best conceived via a few examples (***Plates 9, 10 and 11***) where it can be seen that they range from fairly equant acute angle forms, through to long skinny styles. For convenience this author refers to them as triangular-acute angle textures which covers the majority of shapes observed in two dimensions. They are however *a very powerful texture for recording potential open space* sites and the reader is recommended to inspect the appropriate examples carefully.

◘ PLATE 8

LAYERING, voids, euhedral crystals, small scale triangular textures.

This specimen illustrates crustiform layering in sphalerite and chalcopyrite.

Repeated deposition of sphalerite (dark) and chalcopyrite (yellow) has resulted in a pronounced crustiform-style of layering nucleating from both sides of an original vein-style channelway. Some hints of subhedral sphalerite crystal forms can be seen projecting into the central silica zone (white)

The specimen also exhibits an open vug texture in the central zone. The yellow iron staining surrounding the vug suggests leaching of a pre-existing iron-rich mineral, which is still present in some areas of the specimen (yellow siderite?). The central silica zone is

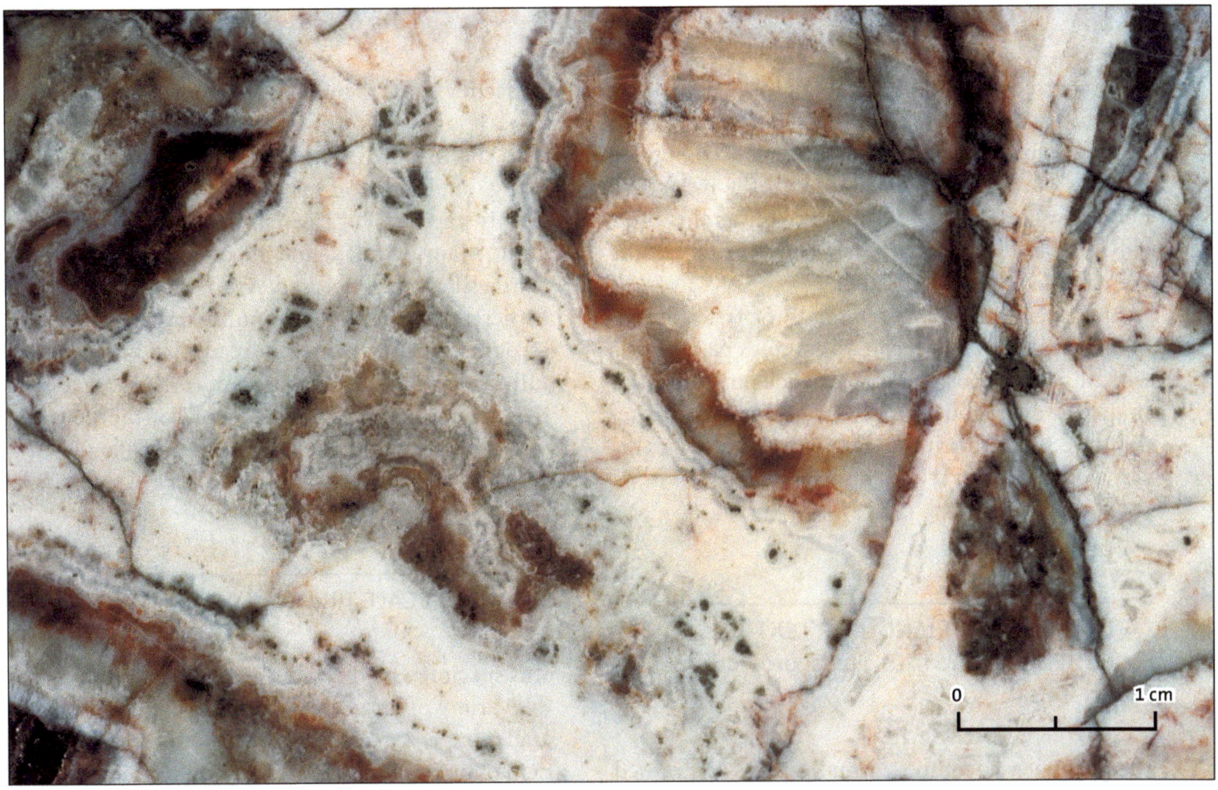

PLATE 7 Layered textures as infill criteria.
Specimen from vein at Quartz Hill gold prospect, near Collinsville, Queensland, Australia.
(Specimen provided by G. Dong).

PLATE 8 Layered textures as infill criteria.
Vein infill from the Cameno (silver, lead, zinc) mine, Mt Albion silver mining district, Queensland, Australia.

more coarsely crystalline and numerous examples of euhedral quartz crystals are visible. Further evidence of infill is provided by the small-scale, "triangular" textures of minerals precipitating in the interstices between quartz crystals. The silver mineral is probably galena, although various silver rich sulphosalts are recorded from this locality.

The specimen is composed totally of various types of infill and obviously records a complex paragenesis with minerals depositing under a wide range of conditions. The quartz phase is actually substantially later than the sphalerite and chalcopyrite, although this is not evident from the plate.

PLATE 9

LARGE SCALE TRIANGULAR TEXTURES, EUHEDRAL CRYSTALS, voids, projecting crystals.

The specimen is composed mainly of fine-grained metasediments which form the framework of the breccia. The fragment interstices are occupied by pyrite, smaller fragments and dark resin which has impregnated during polishing. The pyrite is euhedral to subhedral and projects into the cavities as small crystals.

The rock is part of a hydrothermal intrusive breccia called the Mt Leyshon breccia which is the main host to gold mineralisation. The fragments in this particular sample are mostly rather angular, but significant rounding can be seen on many of the small, interstitial fragments. The interstitial fragments also appear to contain several different rock types.

The alteration history of the breccia is complex and not fully resolved. The most prominent alteration in this specimen is an argillic (white) style which can be seen within fragments and along some fragment rims. It is however difficult to relate this consistently with the infilling phase, in that it is totally absent along some rims and very variable within fragments. At this stage it is regarded as belonging to an earlier event. Many of the fragments contain pyrite and look a little silicified/sericitised. Again this is probably an earlier event. Although the infill component is easy to recognise in this example the reader is advised to carefully acquire a mental picture of the infill shapes as they are much more difficult to see when the incoming mineralisation totally alters the fragments.

PLATE 10

LARGE-SCALE TRIANGULAR TEXTURES.

This specimen has been selected to illustrate some of the variety of larger-scale, semi-triangular to acute angle shaped cavities that occur in breccias. In particular this rock illustrates several examples of the long, skinny, often microvein breed as opposed to the more equant species in ***Plate 9***.

Again the specimen has been specially selected because of the relative lack of wall-rock alteration affecting the fragments. In most breccias the intense alteration of the fragments makes the infill zones very hard to observe, and when the infill mineralogy is similar to the alteration mineralogy the problem becomes compounded. This plate is simply presented to accustom the observer to infill shapes for later use in more difficult situations.

The actual rock is a crackle-style breccia with little fragment rotation. The infilling material is sphalerite (dark), galena (silver) and pyrite (yellow) while the rock fragments are of highly altered (clay, sericite and carbonate) quartz-feldspar porphyry.

PLATE 9 The large scale, triangular acute angle texture. Breccia from Mt Leyshon gold mine, near Charters Towers, Queensland, Australia. This specimen has been selected to illustrate the variety of larger-scale, triangular, semitriangular to acute-angle shaped cavities that form during brecciation. In this case there is relatively limited wall-rock alteration and the form of the permeability is well-defined by the infilling coarse-grained pyrite. (Specimen provided by P. W. Wormald).

PLATE 10 The large scale, triangular-acute angle texture.
Breccia from the Mt Leyshon gold mine near Charters Towers, Queensland, Australia.
(Specimen provided by P. W. Wormald).

PLATE 11

LARGE SCALE TRIANGULAR TEXTURES, edge effects-alteration.

The plate illustrates some of the many forms of triangular-acute angle textures which result from infilling of brecciated rocks. The specimen is specially selected to illustrate the long, skinny, almost vein-like styles, although it also contains more equant varieties.

The rock is composed of brecciated, argillised, fine-grained metasediment (pale) and white clay (kaolin?). The clay component is most probably a hydrothermal precipitate infilling cavities. Although there are no vugs or euhedral crystals it is noted that there are many examples of near-matching fracture walls. Given the degree of brecciation, the general degree of wall-matching strongly suggests a large component of open space with minor alteration.

It is however noted that argillic alteration has been intense in situations where small fragments have become incorporated within the main channelways. This is evidenced by the fact that several small fragments can be seen which have vague boundaries (as opposed to sharp) and others have been almost ghosted-out by intense clay alteration. As noted both in the text and on other plates it is extremely common for the alteration component to be the same as the infilling component, resulting in a blurred edge effect where alteration becomes intense (edge effects).

The infilled channelways range considerably in shape and the specimen has been deliberately selected to allow the observer to become familiar with the range of infill shapes. Although they are obvious here, they are commonly extremely difficult to recognise in very altered rocks or in intrusive breccias where open space is very limited (See ***Plate 19***). The rock also contains a small, grey quartz vein (top left) indicating another phase of hydrothermal brecciation/infill.

2.2.5 Small-Scale Triangular or Partially Triangular Textures

Early formed precipitates show a propensity to develop good crystal shapes which nucleate sporadically on the wall rocks or previously formed crystals. In many instances the later infilling precipitates seem to nucleate more prolifically to form fine-grained products. For instance silica, sericite, clay minerals, chlorite and carbonate are commonly found towards the central zones of cavity infill where they "fill-in" as microcrystalline products. Alternatively the central zone may be infilled by coarsely crystalline material such as fluorite or carbonate, which has sufficient space to achieve crystal form. In either case this produces a wide variety of sharply pointed textures which (with a little artistic license) are commonly referred to as triangular textures. These triangles or V-shaped areas vary enormously in size and form depending upon the arrangement of the crystal substrate and the particular section on view to the observer. They may appear as acute or obtuse angled features, and range from long and thin to short and fat. However they *are a very powerful textural weapon* in seeking out infill in difficult situations. They do however present themselves in a variety of scales and in many different disguises. Numerous examples are depicted throughout the accompanying illustrations with particularly striking occurrences being present in ***Plates 12, 13 and 14.***

PLATE 12

SMALL SCALE TRIANGULAR TEXTURES, euhedral crystals, working backwards.

The specimen has been selected as a very visual example of triangular-acute angle infill texture.

The specimen is composed of dark cassiterite and white quartz. Visually, attention is naturally focussed upon the dark patches, which actually represent cassiterite infilling cavities between/around pre-existing quartz crystals. The later are less obvious but several prominent crystal faces are present. The plate also illustrates the working backwards technique. If a triangular-acute angle texture is suspected, the observer should automatically check to see what it has nucleated on. It is surprising how often this reveals an unnoticed euhedral crystal. The infill shapes vary widely, but attention is drawn to the triangular (star-shaped) forms in the centre mid right of the plate.

The cassiterite in this slide contains two varieties, a dark, early form and a pale, late style which coats and possibly replaces the earlier form. The second, pale cassiterite exhibits some very small-scale and subtle triangular textures as it infills between the earlier, dark cassiterite crystals. This is well-demonstrated towards the right of the star texture mentioned above (centre mid left of plate).

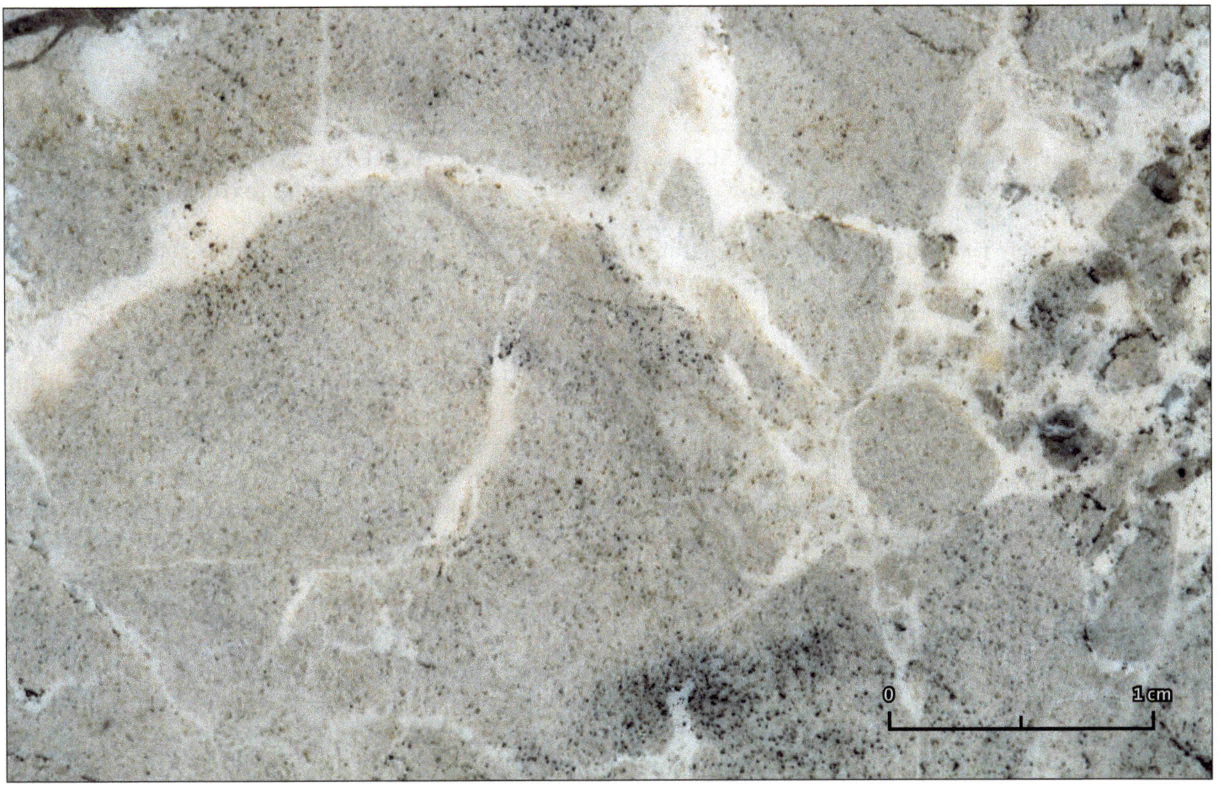

PLATE 11 The large scale, triangular-acute angle texture.
Breccia from Wirralie gold mine, Queensland, Australia.

The small-scale, triangular texture.
PLATE 12 Specimen from the Black King tin-bearing pipe, Herberton, Queensland, Australia.

PLATE 13

LARGE SCALE TRIANGULAR TEXTURES, euhedral crystals.

This specimen has been selected to demonstrate the small triangular-acute angle textures that result from late-phase products precipitating around previously formed crystals. In this case they are highlighted by the pinkish topaz nucleating between tourmaline crystals. It is assumed the specimen is from a vein. Large elongate crystals of tourmaline (dark) are succeded by late pink-yellow topaz which crystallised in the interstices between tourmaline to form a very visual example of a particularly important infill texture the triangular texture.

Quartz has precipitated prior to the tourmaline, and the wall-rocks of fine-grained granite (not visible) have been extensively silicified to form the pale material to the left side of the plate.

The dark-grey sections of the plate to the left are a late, overprinting phase. This is pyrite-rich (now oxidised) and associated with well-developed, siliceous alteration. It is this phase that is responsible for the silica alteration seen in the tourmaline crystals towards the left of the tourmaline zone.

PLATE 14

SMALL SCALE TRIANGULAR TEXTURES, large scale triangular textures, euhedral crystals.

The specimen has been selected as a good example of the triangular-acute angle infill texture involving sulphides.

The right sector contains several very good examples of small scale, triangular-acute angle textures and variously oriented quartz crystals. The sulphides are chalcopyrite (yellow), galena (dark silver) and pyrite (pale yellow). Some larger-scale, triangular-acute angle textures are also present where voids have been infilled by combinations of the above sulphides including various forms of crystalline pyrite. A more subtle variation of the triangular texture is visible to the top right where a late generation of white silica precipitates around pre-existing, clear quartz crystals.

The left sector also contains excellent examples of the key triangular-acute angle texture involving dark-coloured material precipitating around quartz crystals. The darker material consists mostly of dark silica and a silver, acicular mineral (hematite?). An unidentified reddish-brown mineral has precipitated after the hematite and exhibits minute triangular-style textures where it infills between the hematite needles. A small, yellow mineral is also present(?)

The relationship between the upper, sulphide-rich (right) and lower, darker zones (left) is not clear on this specimen. However, it is probable that they represent two different phases of mineralisation. The specimen is composed entirely of infill material, and the complex paragenesis can only be fully unraveled by collecting further specimens from the original location.

PLATE 13 The small-scale, triangular-acute angle texture.
The location of the specimen is unknown, possibly Cooktown, Queensland, Australia.

The small-scale, triangular-acute angle texture.
Specimen location is uncertain, probably the Idarado Mine, San Juan (silver, lead, zinc, gold district) Colorado,
PLATE 14 United States of America.

3 Problems and Working Techniques

3.1 Common Misconceptions

A minor problem encountered by inexperienced observers is the expectation that all infilling crystals will have the same orientation. From most textbook illustrations an expectation is generated that infilling crystals on the side walls of the fissure all appear in a line, somewhat similar to saw-teeth. Whilst this is commonly observed, there are many examples where crystals grow out from the substrate at many different angles. This is especially true for:

(a) Crystals that form rosettes or radiating clusters e.g. quartz, tourmaline, etc.
(b) Crystals that nucleate on fragments.
(c) Crystals that nucleate on the sides of small, irregular void spaces.

In each case the observer will see a variety of euhedral shapes on either cut slabs or rock surfaces. Examples of this feature are especially well illustrated in ***Plate 2***.

A similar misconception is that a precipitating phase will deposit on the substrate as a continuous layer. Unfortunately crystals commonly nucleate irregularly as isolated units or clusters. This is particularly common in breccia style triangular cavities. Examples of this are present in ***Plates 2, 6 and 9***.

A further common problem is the expectation that all infill will appear as crystals. As mentioned previously massive or microcrystalline forms very commonly occur, particularly in the case of silica. (***Plates 3, 6, 7 and 14***)

3.2 Problems with the Well-Formed Crystal Criterion

Although the majority of well-formed crystals observed in hydrothermal environments represent infill textures, there are many exceptions.

Sulphides such as pyrite and arsenopyrite commonly form euhedral shapes via alteration/replacement processes. Carbonates exhibit a similar propensity to grow as good crystal shapes in altered host rocks. Conventional metamorphic/deformational effects will similarly produce good euhedral crystals (porphyroblasts). However with a knowledge of the general geological environment most of these problems are easily resolved.

A major exception to this occurs within the high temperature carbonate alteration systems (skarns). This particular environment produces a wide range of minerals (garnets, pyroxenes, amphiboles, etc) with excellent crystal shapes formed via alteration/replacement processes. It becomes very difficult and commonly impossible to distinguish them from genuine infill.

3.3 The Edge Problem

It is very common for the alteration mineralogy to be the same as that of the infill. This is particularly prevalent within silica-rich systems where quartz infill is accompanied by intense quartz alteration (silicification). Chlorite and sericite infill are similarly commonly linked to intensely chloritised or sericitised zones respectively. In many cases this results in a blurring effect such that the precise location of the side wall of the original channel becomes indistinguishable. Usually the observer can make a good estimate of the side wall position, although this may become difficult with wide zones of monomineralic alteration.

In other cases the problem may be resolved by very careful hand lens inspection which may reveal subtle colour differences between the two similar end-products. It is common with minerals such as chlorite or sericite for the infilling phase is often to be much coarser-grained than the alteration product. Examples of differing degrees of difficulty are shown in ***Plates 1, 11, 15, 16, 17 and 19.***

PLATE 15

EDGE EFFECTS, small scale triangular texture, euhedral minerals, breccia, alteration.

The specimen illustrates the problems that arise in separating infill from alteration. The vital distinction is particularly difficult when the alteration product is identical or similar to the infill minerals (silica). The specimen is composed principally of silica, chalcopyrite, sphalerite and galena and has been selected as a moderately difficult example, illustrating the use of crystals, triangular-acute angle textures, the working backward technique, and alteration effects in breccia recognition.

Identification of infill represents the first step with this specimen. Utilising the crystal shape criterion combined with the triangular-acute angle criterion it can be reasonably interpreted that the region around the chalcopyrite (yellow) represents an infilled open space. The white/grey quartz crystals nucleate from similar, white, coarse-grained silica which is almost certainly the same infilling silica which has not precipitated totally as crystals. This interpretive process is a good example of the working backwards technique.

Some of the darker, triangular-textured zones in the white quartz contain galena/sphalerite, and one also has a late, white (clay?) infill.

The remainder of the specimen consists of dark-grey, siliceous material which contains small dark spots, minor sulphides, and white crystals-triangular textures? These are more difficult to interpret but would be suspected as the results of siliceous alteration of wall rock fragments accompanying the silica-infill precipitation. As alteration intensity increases, the definition of fragments becomes progressively blurred and it is difficult to pinpoint the exact initial boundary of the fragment and the infill channel way. This fragment-blurring is a major problem in breccia recognition. In this specimen the observer is assisted by the fact that the alteration- and infill-silica are different texturally (fine versus coarse) and also in colour (white versus grey). The definition of the original fragment is usually best perceived by holding the specimen out at arms length, which negates the closer-range blurring effect.

3.4 The Working Backwards Technique

Not all infilling minerals precipitate as obviously crystalline products, and silica in particular commonly appears massive or alternatively very fine-grained.

It is good practice to inspect all suspected infill very carefully for the late phase triangular textures previously described. Small triangles of silica, sericite, chlorite, carbonate or clay commonly fill-in the last remnants of "open space". Careful observation of these areas typically reveals suspected crystal shapes protruding into the triangular zone from an apparently massive substrate. Alternatively the triangular texture is such a good indicator that the observer can provisionally assume that it represents infill and that the substrate is probably also infill. By this mechanism two infill phases are identified from the single triangular texture. Numerous examples are given via the illustrations with specific discussion given for ***Plates 1,5, 12, 16, 17 and 19.***

The same philosophy will generally work for isolated crystals which are suspected as infill. It is very probable that anything overgrowing the crystal faces will also be infill (***Plate 5***).

PLATE 16

EDGE EFFECTS, SMALL SCALE TRIANGULAR TEXTURES, working backwards, breccia, alteration.

This specimen has been selected to illustrate problems encountered when the infilling minerals are essentially similar to the alteration minerals. In this case the host sandstone/mudstone has been brecciated and altered to chlorite (darker zones) whilst the infill is also primarily chlorite + silica (dark and white mottled zones). In these circumstances the boundaries between chloritised fragments and chlorite-dominated infill become blurred.

In hand specimen the sample appears uniformly black/dark-green. However on the cut surface considerable textural variation is apparent. Two main textural components are apparent: the fine-grained, dark areas and the dark/white-mottled zones.

The white component of the mottled zones (silica?) exhibits numerous examples of the triangular-acute angle, texture previously illustrated. These range from very convincing through to rather doubtful. However on balance these would be interpreted as infill textures. Utilising the working backwards principle, the

PLATE 15 Separation of alteration from infill (breccia recognition), edge effects.
Specimen from the Isobella (silver, lead, zinc) mine, Herberton, Queensland, Australia.

PLATE 16 Separation of alteration from infill (breccia recognition), edge effects.
Specimen from the Mt Petersen tin deposit, Irvinebank, Queensland, Australia.

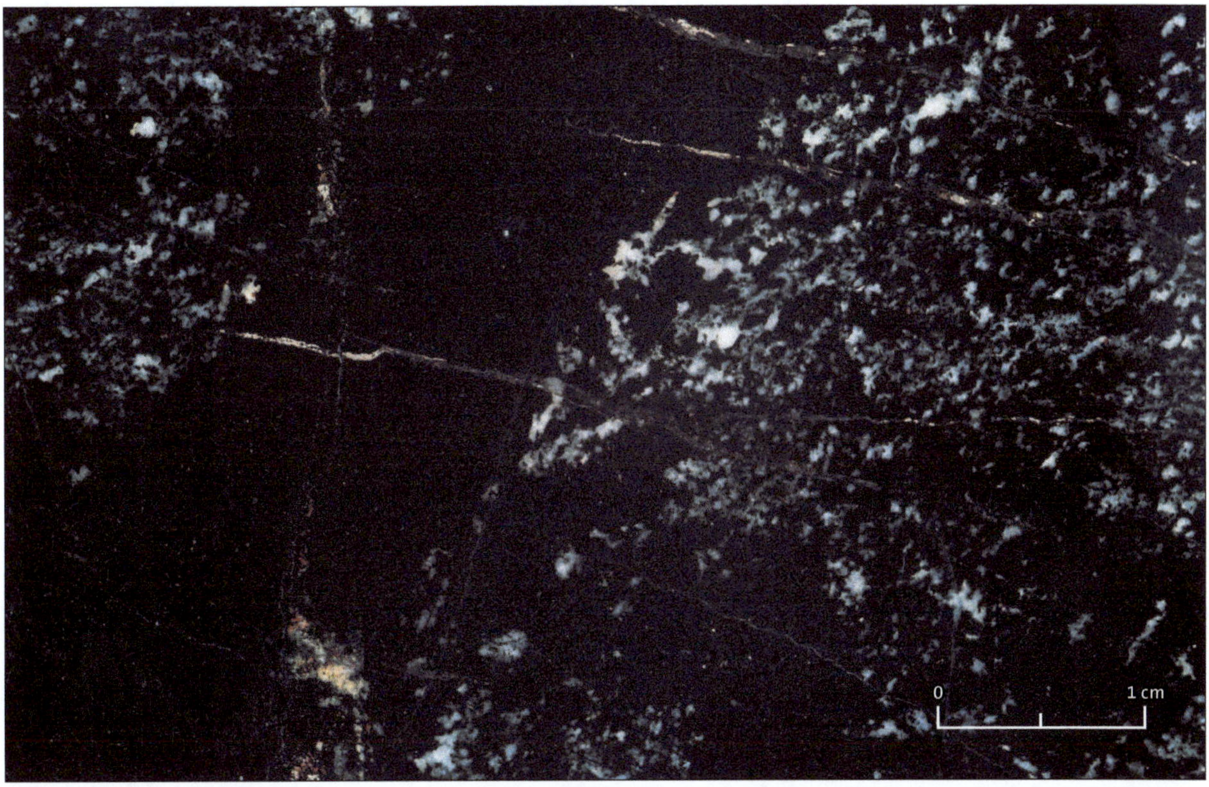

specimen can be further examined to see whether or not the alleged silica infill is overprinting crystalline materials. Chlorite commonly precipitates in cavities as rounded clusters or spikey crystals. Looking behind the quartz patches, there are numerous examples which could be interpreted as rounded clumps (and some spikey zones) and numerous other examples which are less convincing. Again on balance much of the dark material within the mottled zones is probably chlorite infill. The rounded habit of chlorite tends to distort the acute angle texture of subsequent infill and this may account for the "poor quality" of many of the white acute-angle zones. In this instance microscopic examination is probably required to substantiate the above interpretation.

Within the field situation the fine/medium-grained sediments can be seen in all stages of alteration, and there is little doubt that fine, dark zones are chloritised fragments. The large, tabular zone is readily perceived, but the vague, rounded, smaller zones beneath it are less noticeable where the boundaries between the mottled and dark material are rather fuzzy. If the observer stands back a little and instructs the eyes to just distinguish two different things (black and mottled), the dark zones stand out reasonably clearly and can be more easily interpreted as altered fragments.

It is admitted that much of the above is provisional interpretation which requires confirmation via less altered specimens or microscopy. However, it will serve to illustrate the processes involved in looking at specimens where alteration mineralogy is suspected as being the same as that of infill, and also the importance of pulling the two apart to discern the nature of the fluid channelways. Many breccias remain unrecognised due to this problem.

The specimen is cut by some late veinlets (pyrite, silica?, chlorite?, carbonate?)

4 Where and How to Look for Infill Textures – Channelway Recognition

4.1 General Approach

The recognition of infill is aided considerably by applying fundamental geological perceptions which focus the eyes in the right place. The first fundamental rule is to carefully inspect any area of wall rock alteration on the premise that the fluids must have had some means of access. The access is usually via zones of strong permeability which are commonly of "open space" style. Within this context it follows that *attention should focus around the area of maximum alteration.* This concept is easily applied in fracture zones of various kinds and indeed accurate recognition of the open channelways provides valuable information concerning the nature of the fracture zone (tensional, compressional, etc). Breccia systems cause the most problems as the open space permeability is irregularly distributed. An experienced observer will probably suspect breccia from the knobbly nature of the rock which is visible even in highly altered circumstances. The problem then becomes a matter of refining the "eye focussing" to seek the isolated patches of infill. Alternatively, isolated zones of infill may be very obvious within large diffuse zones of alteration, and the process reverses to utilising the infill to gain a perception of the nature of the fragments and matrix components.

4.2 Vein Styles

Vein styles of mineralisation are generally composed of a combination of wall rock alteration and infill. The infill occupies zones of dilatancy resulting from faulting, hydraulic jacking, etc. The distribution of dilatant zones along a fracture surface varies considerably from almost continuous along simple tensional fractures to extremely erratic along fault surfaces. However, a careful field examination of the vein usually reveals zones of significant infill, and most textbooks depict classic examples of crystals growing inwards from the walls in various configurations. In most cases the location of the central fluid channelways and the recognition of the infill zones present very few problems to the observer and need no amplification here. Many examples of vein infill textures are shown via the illustrations.

4.3 Breccia Styles

Infill recognition in brecciated rocks is considerably more difficult, particularly where wall rock alteration is intense. All breccia host rocks contain three major structural components. These include rocks (fragments), finer-scale rock fragments and powders (matrix), and void space (potential infill sites). The amount of void space varies enormously from around 20-50 % in collapse breccias to very minor in milled breccias. Although the void spaces within a breccia may exhibit a wide range of size and shape, they tend to be vaguely triangular in cross section with sharp acute angles existing between fragments. *This triangular texture is a prime criterion enabling breccia and infill recognition* even within very altered rocks. Within this context breccias exhibit patches of infill which on the rock or slabbed face do not appear to join up. Within the broad scale triangular texture it is commonly possible to identify small-scale triangular textures of infill around crystals as described in the previous section concerning infill identification criteria. Good examples are depicted in ***Plates 9, 10, 11*** and recognition of infill in highly altered breccia systems is discussed via ***Plates 16, 17, 18 and 19.***

4.4 Other Styles

4.4.1 Miarolitic

Exploration geologists concerned with magmatic-related deposits have become increasingly aware of indicators which might represent clues to potentially ore-bearing/ore-producing plutons. The concepts are best understood within the tin and tungsten systems, but have been steadily extending to find applications within pegmatite, molybdenum, copper and more recently gold systems. These involve the recognition of features considered to have resulted from interaction between magmatic and hydrothermal phases. They include various species of miarolitic textures, granophyric intergrowths, apical pegmatites (stockscheiders), unidirectional solidification textures (UST's), pegmatitic granites and background alteration in granitoids.

The miarolitic/pegmatitic texture is of considerable interest, occurring very commonly within the apical zones of ore-related plutons and also within their dyke/sill style derivatives. (*Plate 20*)

In most systems the miarolitic/pegmatitic facies is unmineralised and seems to have formed slightly earlier than the main introduction of hydrothermal fluids. However in rare cases the infill-like textures do contain valuable hydrothermal precipitates, resulting in large low-grade, "disseminated" style mineralisation (*Plate 21*). Opinions differ as to how much of the apparent infilling is of magmatic versus hydrothermal origin.

◘ PLATE 17

EDGE EFFECTS, SMALL SCALE TRIANGULAR TEXTURE, euhedral crystals, working backwards, alteration, breccia.

The specimen shows several features which raise suspicion concerning potential infill towards the right hand side. Firstly, the white quartz displays vague hints of crystalline textures and vague hints of vugs. Although not visible here, the reverse side of the rock displays several obvious quartz crystals in vugs. Secondly, the dark mineral (tourmaline) attracts attention with vague hints of triangular-acute angle textures and good crystalline shapes. Utilising the working backwards technique it can be observed that the tourmaline substrate is coarse-grained white mica (sericite) growing in clustered spikey or slightly rounded aggregates. The tourmaline is precipitated after the sericite, and the nature of the mica-sericite distorts the normal acute-angle texture to a more muted, rounded form. The paragenesis appears to be quartz followed by white-mica and then tourmaline.

The left hand side of the photograph is dominated by white mica and silica. However, the white mica is generally very fine-grained and the grey silica is poorly defined. The sericite/silica boundary is also rather blurred. This zone represents sericitised granite where the feldspars and biotite have become extensively sericitised, whilst the granite-quartz remains but is under attack around the margins.

Whilst it is relatively easy to pick the infill zone from the alteration zone, the exact border between the two is blurred. It is worth noting again that when the infill and alteration minerals are similar, the infill is commonly coarser-grained.

◘ PLATE 18

SMALL SCALE TRIANGULAR TEXTURES, euhedral crystals, working backwards alteration, breccia.

The specimen consists of deformed (sheared) granite which has been extensively chloritised (alteration of feldspars and biotite via microfractures).

Within the general chloritisation there are many examples of bright red material (iron-oxide) which upon close inspection can be seen to form a semi-linking, anastomosing network across the specimen. At one point (top right) the red material exhibits excellent triangular-acute angle textures and is clearly infillings around quartz crystals. The quartz crystals have a tendency to be whiter than those of the altered granite.

Similar small vugs (not visible here) occur throughout the hand specimen and there is no doubt that the rock has been substantially broken (brittle fracture) prior to the introduction of quartz and hematite.

It is probable that the chloritic alteration is associated with an earlier, ductile-shearing event whilst the silica ± red material has been subsequently introduced during major brittle-fracturing, brecciation.

The main point of this example is not to identify breccia types, but to illustrate that there is a hidden breccia within the altered rock. This can be identified via infill textures but is far from obvious in the field or on the slab. Without the red colouration the recognition of isolated vugs would be even more difficult!

PLATE 17 Distinction of alteration from infill – breccia recognition
Specimen from the Southern Cross tin mine, Herberton, Queensland, Australia.

Breccia recognition.
Specimen from the breccia pipe system at Ardlethan tin mine, Ardwest ore zone, New South Wales, Australia.
PLATE 18 (Specimen provided by G. W. Clarke).

◘ PLATE 19

LARGE SCALE TRIANGULAR TEXTURE, EDGE EFFECTS, working backwards, alteration, breccia.

The specimen has been selected to illustrate the difficulty and importance of infill recognition in intrusive breccia systems. The Ardlethan intrusive breccia systems remained unrecognised throughout some ten years of extraction, and their true nature only became apparent during the latter years of mine life.

The specimen represents one small area of an intrusive breccia pipe which has been etched with hydrofluoric acid in an attempt to facilitate textural recognition. The key feature of the illustration concerns the narrow discontinuous black/brown zones. These are composed of tourmaline (black), cassiterite? (brown) and silica (grey). More importantly, their curving and extended acute-angle vaguely triangular shapes are very suggestive of infill textures (See *Plates 10 and 11* for the elongate triangular-acute angle style).

The remainder of the rock consists of two forms of highly altered, sericitised granite. The first form is as fragments which can be vaguely discerned by locating coarse, white patches associated with coarse, clear quartz. The white represents sericitic/argillic alteration of feldspar whilst the quartz is original granite quartz. The general granitic texture is retained.

The remainder represents well-milled granite and is characterised by the general lack of coarse, white patches (ex-feldspar), and the general lack of coarse-grained quartz as an interlocking mosaic. It represents milled/altered material derived from the original granite host.

Not surprisingly, the distinction between fragments and milled matrix is commonly difficult, and is in fact better observed on larger specimens. However the key factor here is infill recognition. It is usual for large-scale breccia to consist of disconnected open spaces which subsequently become infilled, and the suspicions of infill given by the discontinuous, curving, dark zones would automatically invoke a suspicion of breccia. The small-scale nature of the dark zones would also arouse suspicions of intrusive brecciation where void space tends to be minimised due to the dusty, milled matrix filling most of the space between fragments. It was the recognition of the above, potential infill textures that finally led research workers to understand the nature and origin of the Ardlethan system.

◘ PLATE 20

EUHEDRAL CRYSTALS, SMALL SCALE TRIANGULAR EFFECTS.

This form of bi-textured rock is particularly common in tin and tungsten mineral districts, usually in association with late-stage, highly fractionated, mineralising granitoids. The granitoids and associated rocks are commonly geochemically highly anomalous with potentially high levels of Sn, W, F, B, Rb, etc. The rock type is particularly common in the apical zones of small intrusions (and associated sills/dykes) which are directly associated with mineralisation. It may link to a coarser, pegmatoidal facies, a marginal pegmatite zone (stockscheider) or UST zones (Unidirectional Solidification Textures). All of the above are generally interpreted as resulting from crystallisation of magma in direct association with hydrothermal components. Similar textures have been reported from pegmatite, tantalum, molybdenum and gold environments.

The rock consists of two components: a background of fine-grained, granophyric granite and an interconnecting, anastomosing network of coarser-grained patches. The latter are composed of quartz-feldspar and biotite. The quartz and feldspar exhibit good crystal form whilst the biotite exhibits good triangular-acute angle texture as it precipitated between the quartz and feldspar crystals. This clearly qualifies as an infill texture although opinion differs as to whether it results from hydrothermal, magmatic, or some kind of transitional fluid. The finer-grained matrix component has a grey tinge and may have been mildly, hydrothermally altered (sericitised?). The rock is not a breccia and the texture is of primary origin. Although not depicted here many of these rocks contain actual void spaces in the central zones of the coarsely crystalline infill component.

PLATE 19 Breccia recognition.
Specimen from the Ardwest-Wild Cherry intrusive breccia pipe system, Ardlethan tin mine, New South Wales, Australia.
(Specimen provided by G.W. Clarke).

Miarolitic granite and infill textures.
Specimen from Herberton, Queensland, Australia.
PLATE 20 A typical specimen of miarolitic granite (also referred to as pegmatitic granite).

PLATE 21

EUHEDRAL CRYSTALS, VOIDS, PROJECTING CRYSTALS, small scale triangular textures, edge effects, alteration.

The specimen has been selected to illustrate a mineralised example of miarolitic granite.

The specimen consists of two textural zones, a green-grey, fine-grained component and a coarser-grained, pink-black component. The fine-grained component is moderately chloritised, granophyric granite (chloritised feldspars) whilst the pink-black component is primarily feldspar (pink), silica (white) and cassiterite (black/brown).

The cassiterite nucleates upon quartz and/or feldspar and either fills the central zone, or forms crystals projecting inwards. In either instance the nucleating regions exhibit numerous examples of excellent triangular-acute angle textures (cassiterite crystal faces are more obvious upon uncut surfaces).

Other minerals within the central zone include fluorite, sericite, pyrite, and arsenopyrite. These are not all present within the photograph although fine yellow sericite can be observed as late infill with triangular-acute angle textures.

This rock is directly analagous to the unmineralised miarolitic granite (*Plate 20*). The important differences are that the fine-grained ground mass is highly altered, and the coarse-grained component shows ample evidence of infill by hydrothermal minerals (crystals, triangular textures) such as sericite, cassiterite, fluorite, sulphides, etc.

It can be interpreted that the miarolitic rock has been the host to a hydrothermal fluid via the porosity/permeability represented by the cavities now filled by the coarse-grained component. The fluid may have been generated around the time of igneous crystallisation, or may have arrived subsequently. The rock is not a breccia, but does represent a porous/permeable host.

The disseminated deposits are flat lying zones within the general miarolitic granite although it has been suggested (P. J. Pollard, pers. comm.) that they represent areas where the percentage of potential void space is higher than usual. Many deposits labeled greisen or disseminated may have an origin of this nature, and the key recognition feature is separating infill from alteration.

4.4.2 Igneous Interstitial Mineralisation

It has been recently recognised that many granites evolve hydrothermal fluids at a very late stage of crystallisation. These fluids appear to become trapped within the pluton and either occupy small void spaces between the previously crystallised, magmatic minerals or infiltrate along grain boundaries to produce widespread "background" alteration. The minute interstitial cavities are not usually visible by eye, but can be clearly seen beneath the microscope. The infilling hydrothermal minerals exhibit all the characteristics described from larger examples within this text. A relatively coarse-grained example of this process is present at the Zaaiplaats tin mine, South Africa where the infill minerals produce ore grade horizons, and the triangular textures are commonly visible via hand lens examination (*Plate 22*).

4.4.3 Dissolution within Granite Systems

Several well-documented examples have been reported where granitoid wall-rocks have their quartz selectively dissolved to create sponge-textured rocks. The sponge holes become totally infilled and the net effect is of numerous small, isolated spots of infill within otherwise solid rock (*Plate 23*). This can be very difficult to recognise, especially when the alteration of the rock is mineralogically similar to that of the infilling crystals. For instance, the phenomenon is common in feldspathised rocks such that albitic alteration may be accompanied by albite crystal infill. Similar examples have been noted with potassium feldspars, and also silica.

PLATE 22

SMALL SCALE TRIANGULAR TEXTURES, EUHEDRAL CRYSTAL FACES

These specimens have been selected to illustrate an unusual form of infill which escaped detection for over fifty years on account of the small scale.

The plate shows apparently unaltered granite composed of red feldspars, quartz and minor dark minerals (biotite). However the very red character, the lack of plagioclase, and the dull lustre of the biotite might raise suspicions of alteration. Under the microscope the granite displays many alteration features the most noticeable of which are that all the biotite has been altered to chlorite and the feldspars

PLATE 21 Miarolitic granite and infill textures.
Specimen from disseminated ore zone at Zaaiplaats tin mine, Limpopo, South Africa.
(Specimen provided by P. J. Pollard).

Micro-infill textures in granite.
PLATE 22 Specimens from the disseminated ore zone at Zaaiplaats tin mine, Limpopo, South Africa.

are complex mixtures of K-feldspar and albite.

The microscope also reveals very tiny, triangle-acute angle textures which are interstitial to the original granite minerals. These minute features contain a suite of hydrothermal minerals which exhibit crystalline features, and internal triangular textures. They are miniature examples of the infill textures described throughout this publication and include feldspar, quartz, cassiterite, scheelite, synchisite, sericite, chlorite, fluorite, hematite, calcite, etc.

On the hand specimen (left hand side) their presence is just visible via the regions of pale-yellow sericite. A bright-yellow, triangular-textured patch is clearly visible and several others are present which are slightly more obscure due to much of the sericite washing out during sample preparation. Similar patches of dark-purple fluorite are present.

The larger scale inset plate illustrates a good example of an interstitial vug. The key feature is the dark brown cassiterite which occupies the centre. This exhibits excellent triangular-acute angle texture and utilising the working backwards techniques it can be seen to overgrow crystals of quartz (clear-white) and a crystal of scheelite (whiter) opposite the quartz crystals. Purple fluorite is also visible.

The rock is not breccia, and the ore appears to have been formed by hydrothermal fluids precipitating within interstitial space between the granite minerals. The ore zone is essentially flat and is mined via an open cut. Ore reserves are ill-defined, but several million tons of low grade tin-tungsten are are present. This style of mineralisation has only been documented at Zaaiplaats, but there seems no reason to prevent similar occurrences within other granite-related elements systems (e. g., molybdenum, copper, gold etc). The key to their origin is again given by recognition of infill textures.

◘ PLATE 23

EUHEDRAL CRYSTALS, VOIDS, PROJECTING CRYSTALS, small scale triangular textures.

The specimen indicates an unusual situation where cavities have resulted from dissolution of quartz within granite.

The specimen covers the extreme edge of Maggs Pipe at the changeover from altered granite to one of the unusual Zaaiplaats high-grade pipes.

The mineralisation occurs as cassiterite together with chlorite (both dark coloured). The cassiterite crystals are just visible in places via some small reflective surfaces. The cassiterite is predominately crystalline and was deposited either on crystalline chlorite or directly upon the granite feldspars. Many of the cassiterite-chlorite patches are actually open vugs with inwardly projecting cassiterite crystals and rare chlorite clusters. They contain obvious characteristics of infill (cavities, crystal faces and triangular-acute angle textures against feldspar crystals). The granite feldspars exhibit mild chlorite alteration adjacent to the chlorite-cassiterite patches.

The rock is not brecciated and it is very noticeable that as the infill zones appear, the granite quartz vanishes. At the pipe edge the cassiterite-chlorite patches mimic the shapes of the granite quartz.

It is concluded that something has dissolved the granite quartz, thus creating the porosity/permeability required for the hydrothermal deposition. This phenomenon has recently been noted in a range of tin-bearing rocks, especially those involved with albitisation and is also present in K-feldspar alteration associated with the lower levels of quartz-wolframite vein systems (Xihuashan mine, China). The nature of the fluids responsible is unknown although strong alkalinity is one possible candidate. The dissolution effect has critical implications for both fluid compositions and in the creation of permeable channelways, and may well be more widespread than currently recognised. It is very difficult to recognise without a good understanding and recognition of the infill (and alteration) process.

PLATE 23 Euhedral crystals, "small scale" triangular textures. Dissolution and infill. Specimen from edge of Maggs Pipe, Zaaiplaats tin mine, South Africa. Photograph by Kath Hotschilt Studios, Townsville, Queensland, Australia.

Volume 2 – Alteration Textures

Volume Front Page

Specimen from Mt Cobalt region, Queensland, Australia. This specimen has been placed on the cover because it looks very attractive. It does also illustrate hematite alteration adjacent to veins containing an infill of cobaltite (white) and quartz (dark).

◘ Plates Volume 2 • ALTERATION TEXTURES

1 Introduction

Recognition of the effects of hydrothermal fluids upon their host rocks is one of the most important and fundamental skills required by geologists involved in ore search. Naturally recognition needs to be followed by comprehension of what the alteration may mean, but the initial observation remains crucial to any subsequent interpretation.

Most of our teaching institutions can only devote a small proportion of their curriculum to introduce students to the many aspects of this difficult subject, and it is not surprising that most geologists emerge with some rather hazy concepts and substantial insecurities concerning their recognition skills. The author can still remember such minor problems as:

- walking over a kilometre of potassic alteration without even noticing;
- working with mineralised granites for some ten years without perceiving their subtle selective-pervasive alteration characteristics;
- nodding wisely during mentions of advanced argillic or propylitic alteration whilst quietly thinking “I wonder what they look like?”

The terminology of alteration is indeed daunting. A short list of types could include such delights as silicification, sericitisation, chloritisation, K-feldspathisation, albitisation, hematisation, biotitisation, tourmalinisation, argillisation, sulphidation, topazisation, dolomitisation or even scapolitisation!

Understanding alteration ultimately incorporates a range of skills which involve recognition of general alteration minerals, structural styles, mineralogical changes, chemical changes, mechanisms of fluid access, paragenesis, zonal distribution, and a knowledge of how all of these relate to a particular style of mineralisation. It would be a little difficult to cover all of these in one small volume and consequently this volume concentrates upon the recognition of the more common types, whilst reviewing some of the general principles and thought processes involved. The focus is upon the mineralogical and textural features that can be seen by eye or with a hand lens. The text is written for the beginner, but hopefully will be of general assistance to all. The third volume looks at the difficulties involved in unravelling overprinting-paragenesis, and systems interpretation. This volume also deliberately concentrates on alteration in aluminosilicate systems and thus conveniently excludes the more complex area of alteration within carbonates (skarn, carbonate replacement). The text is not intended as a full scale alteration review. However, for general interest it could be read in conjunction with the relevant sections of Guilbert and Park (1986).

2 Basic Principles

It is not intended to discuss the details of alteration theory within this text, but rather to concentrate upon the recognition process. However, it is perhaps worth a reminder that the term refers to the effects that a hydrothermal fluid imparts upon a host rock. These take the form of mineralogical and/or textural changes which are referred to by a variety of terms (alteration, wall-rock alteration, replacement, and metasomatism). The hydrothermal fluid is normally channelled through the rock by either primary (pore space) or secondary porosity (fractures) and in general terms the degree of alteration increases as the main channelway is approached. The subject is of priority interest to economic geologists as the altered rocks form natural halos adjacent to valuable ores, as well as providing a host of valuable clues concerning the composition and physical parameters of the ore fluids. The alteration effects range from small selvedges adjacent to cracks up to kilometre scale zones surrounding breccia pipes or porphyry systems.

Plate 1 is intended to introduce the general concept of alteration in a simple form, and depicts a quartz vein in granite with an irregular dark halo of alteration separating the vein from the host rock. It is worth covering a few points which are occasionally confused.

It is generally assumed (although not necessarily true) that the central vein marks the position of an originally open channelway which formed the conduit for moving hydrothermal fluid. The fluid is assumed to have reacted with the minerals of the wall rock to create a new assemblage.

The amount of change is more intense closest to the channelway and the alteration process is adjudged to continue until chemical or physical conditions change. Alteration is thus conceived as a moving front and it is quite common for the assemblage adjacent to the vein to be quite different to that further away. Similarly, textural changes are normally most intense adjacent to the channel. As the process continues the size of the halo increases and the inner zones grow outwards to overprint the initially weaker outer alteration. A new outer zone forms as fresh rock is encountered by the expanding front.

It can seen from the example (***Plate 1***) that the degree of development is quite variable. The method by which the fluid moves through the rock is commonly uncertain, and arguably represents various combinations of diffusion, grain boundary controls, and microfracturing. The factors causing/controlling the reactions are not the subject of this text but there are many potential variables including temperature, pressure, pH, Eh, fluid composition, wall rock composition, rate of flow and periodicity of flow. Indeed one of the major scientific reasons for studying alteration is to try and deduce as much as possible concerning the ore fluid parameters.

A most important (and generally neglected) facet of wall rock alteration studies is to realise that in most instances the alteration implies the presence of a fluid channelway and it is most important to locate and understand the channelway. This is commonly a fluid-filled (open space) zone which is marked by infill textures. The latter may be difficult to see, but anyone who observes alteration *should automatically look for the infill corollary.* In this instance (***Plate 1***) the quartz vein is presumed to be the infill component and displays both void texture and obvious quartz crystal (comb quartz) layers.

This raises the nasty question – was the channelway filled at the same time that the alteration occurred? In most cases this question is conveniently (or conventionally) ignored and it is assumed to be the case! Presumably as infill proceeds fluid flow must eventually be constricted or cease. The presumption of contemporaneity is supported in some 60-80 % of cases where the infilling components are similar to those occurring as alteration products. It can also be noted that the style of channelway has something to say about the structural environment. The significant gap comprising the channel could be interpreted as a tensional feature and/or a product of hydrothermal jacking.

The gradational nature of most alteration zones provides an opportunity to observe which minerals within the host rock are the most susceptible to reaction and consequently represents an area worthy of special alteration. In this example (***Plate 1***) the feldspar component is replaced by a buff-grey mineral (fine -grained white mica-loosely termed sericite) and there is just a vague hint of silica increase also at the expense of feldspar. The alteration is texturally retentive.

Another important factor to constantly realise is

PLATE 1 **SERICITIC ALTERATION**
Quartz vein with dark (sericite-dominated) alteration halo in fine granite host.
New St. Patrick copper mine, Copper Firing Line, Herberton, Queensland, Australia.

that chemical changes have obviously occurred and these should be mentally recorded at the observational stage. Some elements have been added to the rock whilst others have been removed. The elements added are reflected in the new mineralogy, and those removed are presumed to have gone into solution and passed on with the fluid. In this sense a hydrothermal fluid is constantly modifying in composition as a result of alteration along its path. In this instance (*Plate 1*) the obvious visual change involves the addition of a probably potassium-rich mica in place of potassium-rich K-feldspars and minor albite. Given this it would be difficult to make a definitive conclusion concerning potassium adjustment although there must be an increase in (OH) components. If K-feldspar predominates in the original rock there may be an overall potassium loss. There is no substitute for careful observation and the author strongly recommends that each specimen suspected as alteration be treated from first principles. The answers to the following questions should be considered:

Question I	**Does alteration exist within the specimens being examined? (General alteration recognition)**
Question II	**Does the observer need to move to ultimately comprehend the full picture? (Observational positioning)**
Question III	**What are the alteration minerals present? (Mineral identification)**
Question IV	**How was the host rock altered on a detailed mineral by mineral basis? What was altered to what? (Changeover zone observation)**
Question V	**What were the textural changes? Was alteration texturally retentive or texturally destructive? (Textural observation)**
Question VI	**How did the fluids gaining access? (Channelway identification)**
Question VII	**Were elements added or subtracted as a result of alteration? (Chemical awareness)**

3 Approach to Alteration

3.1 Alteration Recognition

Is the rock altered?

It is a reasonable question to ask – "How does one recognise an altered rock in the first place?" This initial recognition step ranges from very easy to extremely difficult. The more difficult cases are not the subject of this text, but the beginner can rest assured that no geologist is fully competent in the many facets of alteration recognition. It is only in recent years that geologists have begun to suspect that the fine-grained host rocks to the Mt Isa mineralisation have been altered on a grand scale. Similarly the alteration effects produced by the large scale migration of mineralising basinal brines are extremely difficult to separate from more normal diagenetic changes and the potential effects of regional metamorphic alteration. Further complications can arise from the fact that higher grades of metamorphism induce major changes upon pre-existing alteration assemblages, a situation common within the exhalative and volcanogenic deposit styles. Fortunately most alteration around fluid conduits is not heavily disguised and alteration is quickly suspected by:-

(1) Halos adjacent to vein style mineralisation (***Plates 1, 1A, 3, 9, 16***).
(2) The presence of partially altered (sick looking!) rocks either in or around mineralisation. These commonly retain their original textures and are obviously altered derivatives of the host rocks (***Plates 8, 13, 17, 22***).
(3) An association of mineralisation with mineral assemblages that immediately come under suspicion as being common hydrothermal alteration minerals (Section 5). From this perspective it obviously becomes a top priority to be able to recognise common alteration products (***Plates 2C, 10, 13, 14, 17, 19, 21***).
(4) The presence of any of the above in conjunction with obvious infill textures (***Plates 4, 6, 13, 16, 20***). Alteration is the natural corollary of infill, and observers are recommended to acquaint themselves with both the common and more subtle features of infill textures.

It is worth noting that the distinction between alteration and infill is a common source of error in both applied and academic studies. Numerous paragenetic tables are published annually where the two components are mixed together! This text is written to introduce the observer to alteration study and most of the examples depict relatively straight forward situations which can be used to build observational confidence.

3.2 The Movement Principle – Observational Positioning

What is the host rock?

Nearly all the problems concerning alteration can be removed by adopting a basic approach which constitutes the first rule of alteration observation. This is best expressed by the simple word – MOVE. If a rock is perceived to be altered either move the eyes or if necessary the entire body to a point which is considered to be unaltered. This process of *observational positioning* can be restated as – *find the host rock*. This may sound somewhat trite, but it is extremely common for people to try and sort out what has happened by standing in the middle of an alteration zone where all traces of the original rock have been destroyed. This usually occurs because it is also the site of mineralisation which naturally attracts attention. Most of the examples utilised in this text are taken from vein controlled situations which only require an eyeball flicker. However, larger alteration zones associated with intensely shattered or brecciated rocks, or permeable sediments may require movement ranging from the metre to kilometre scales.

3.3 Mineral Identification – Nomenclature

The problem of nomenclature causes immense problems to beginners who are understandably anxious to come up with a name. Alteration terminology is seeped with tradition, and generic style names such as propylitic, phyllic, potassic and greisen are entrenched. These are best avoided in the initial stages of recognition. Names such as chloritisation and silicification are a little more comprehensible being generally derived by nominating the dominant mineral. However, even these terms create confusion for the beginner who very quickly notes that many alteration assemblages contain two or three very obvious minerals and is uncertain which one to nominate for a general title. By far the best procedure at the early observation stage is to nominate all the visible minerals and use the assemblage as a name. Thus a mixture of quartz, chlorite, calcite is called a quartz, chlorite, calcite alteration assemblage. Utilising this approach ensures that everyone is clear what is being talked about. A broader name can be applied once this has been clearly established, and the observer is effectively free from all the problems that abound from the very confused nomenclature. A much more fundamental problem arises with both experienced and novice observers who have difficulties with identifying fine-grained alteration products. Hopefully the plates within this text will be of some assistance. However, if field recognition is a problem it is wise to simply record that the new assemblage is composed of a red mineral, a green one, and a grey one or whatever is appropriate. The important thing is to *look*. A thin section will solve the problem for the beginner and as experience is gained the detailed "looking" will quickly convert into confident recognition. In all of this a hand lens inspection is vital. In all cases it is recommended to look in detail at the rock before and after any thin section assistance. This process radically reinforces the observational learning curve.

3.4 Changeover Observation (Transitional zones)

How are the new minerals derived from the host rock?

The most important observational zone is the point of change between the host and the altered rocks. This is affectionately termed as looking for the half/half rocks, or zones where the host rocks are only partially altered. This zone usually retains the textural character of the original rock, and it is possible to see the original mineral being converted into new ones. With coarser-grained rocks it can usually be seen that specific minerals behave differently. Feldspars and ferromagnesian minerals are very prone to early change whilst quartz is usually more resistant. From this simple approach it is usually possible to see "What is going on".

Obviously it helps if the observer is good at recognising alteration minerals, but even at the elementary stage it is possible to say that the plagioclase has turned pink, the biotite has gone green while quartz has been unaffected.

What is the mineralogy of the alteration assemblage – What will I call it?

3.5 Textural Observation

What are the textural changes involved in the alteration process?

Most alteration assemblages exhibit an incredible degree of textural inheritance from their host rocks. Coarse-grained rocks such as granites or gneisses produce coarse-grained alteration products, whereas fine-grained rocks such as micro granites or shales change to fine-grained products. Similarly coarse- or fine-grained layering or original layers of different composition usually retain some form of textural similarity between host rocks and alteration products. This is brought about by the general resistance of quartz to change. Thus granite quartz, quartz phenocrysts and sedimentary quartz are commonly the last to convert. Consequently even quite strongly altered rocks are still texturally coherent with their hosts. Obviously textural retention is at a minimum within the most intensely altered zones and this again emphasises the first rule of alteration, which is work from the known towards the unknown and carefully note what happens. Some alteration styles are noted for being particularly texturally destructive and the observer will find examples where it is impossible to see any obvious textural retention. It is a little unfortunate that two very common alteration styles (silica and sulphide alteration/replacement) fall into this category.

3.6 Channelway Identification

How did the fluids gain access?

This is a critically important question which must be asked on every occasion where alteration is suspected. Alteration is often taught, and almost always illustrated as spreading out from a central fluid channelway. This vein style approach is very convenient as an introduction to the subject, and is the basis of most of the illustrations used here. Most people have little trouble with this concept although they do experience practical problems in sorting out the infill component from the alteration effects. However, there are many other styles of access which create confusion. One of the most common is via various forms of brecciated material. In these instances the fluids permeate through the breccia in a relatively irregular way utilising combinations of open spaces and rock flour as the major channelways. Tracking and comprehending alteration in sizeable breccia zones requires well-developed skills at recognising both infill and alteration textures. Ore fluids also exhibit a habit o(being very layer selective. This occurs in response to an individual horizon being either very permeable or very chemically reactive. The combination of permeability - reactivity is highly attractive and can produce layer controlled alteration over many tens of metres (or even greater distances). Poorly consolidated sediments or sediments with secondary induced porosity are prime targets and chemical reaction may be encouraged by the presence of carbonate, iron-rich, felspathic, or carbonaceous components. Fracturing and infill are often absent or difficult to detect and if such alteration is suspected the best course of action is to move laterally to locate the potential host and changeover point. Usually the problem can be solved on the metre scale, but with grand scale examples such as roll front uranium systems or migrating Pb-Zn basinal brines the observer may have a long walk!

Other less apparent channelways occur in relation to subtle background semi-pervasive alteration zones within granitoids. These range from obvious miarolitic cavities to much more subtle microscopic interstitial cavities, cracks, and grain boundary controls. In many of these cases the fluids probably emerge at the very last stages of magmatic crystallisation. It should be obvious from the above that every physical and mental effort should be made to understand the method of fluid access (channelways). The access recognition problem increases with the scale and/or intensity of the alteration effect. This volume is aimed at recognising and characterising alteration in the first instances, and can only hope to raise the level of observer awareness in more complex situations. However, at least one example of most channelway styles has been included, and a full discussion included with the figure captions.

3.7 Chemical Awareness

Are there any obvious chemical changes?

This volume has not been written from the chemical viewpoint and will no doubt incur minor irritation from fluid geochemists. The chemical adjustments are of course of paramount importance in gaining information as to the nature of the ore fluids. This field of endeavour is of equal importance to that of recognition/observation and it is a little unfortunate that many of the top observers are very poor fluid geochemists and visa versa. However, even the chemically disadvantaged should at least cultivate the habit of noting obvious chemical parameters. This requires an elementary knowledge of the chemical composition of common alteration minerals. It is a simple operation to deduce that a quartzite which has been converted to chlorite has lost silica and gained magnesium, iron and aluminium. In most cases the gross chemical change due to alteration can be readily derived from the obvious differences between host and alteration mineralogy. Some details of common alteration minerals are given in Section 4, and anyone aspiring to alteration expertise should ensure they are conversant with the essential mineral chemistry

3.8 Examples of Alteration Types and Structural Style

The following plates (*1A–22*) are included to give the reader a general visual impression of various alteration types within a variety of structural styles. In each case a detailed description has been provided which incorporates the thought procedure outlined above. It is strongly recommended that the beginner go over each one with some care. Detailed observation is the main habit which is worth cultivating.

PLATE 1A

SERICITIC ALTERATION

Quartz vein with dark (sericite-dominated) alteration halo in fine granite host. New St. Patrick copper mine, Copper Firing Line, Herberton, Queensland, Australia.

This plate is presented to give the reader a visual impression of the very common sericitic alteration type. It has been included twice (***Plate 1 and Plate 1A***) to facilitate both the discussion in Section 2, and the complete description here.

GENERAL ALTERATION RECOGNITION

Alteration is automatically anticipated from the presence of the dominant dark halo adjacent to the quartz vein.

OBSERVATIONAL POSITIONING

Application of the "MOVE principle" only requires simple eyeball movement traversing from the relatively unaltered granite towards the central vein style (white quartz) channelway.

MINERAL IDENTIFICATION

The new alteration assemblage is composed of fine-grained dark grey-green-buff material, together with a coarser grained dark brown-black mineral and a grey-white vitreous mineral.

The latter two obviously relate to similar minerals in the host rock (dark biotite and grey-white quartz). The dominant grey-green-buff mineral is too fine-grained for positive identification. However, this colour is very typical of fine-grained white mica traditionally given the vague general name *sericite*. The term horrifies mineralogists owing to its imprecise nature. However, it is totally ingrained in common geological parlance as a useful general term. Most sericite turns out to be various species of muscovite but usually requires X-ray Diffraction and electron microprobe examination for detailed clarification. These procedures are time consuming, costly, require advanced research expertise, and are not usually pursued at the exploration level of investigation. The beginner should note that fine-grained sericite is frequently much darker in colour than might be expected. Close inspection of ***Plate 1A*** also leaves the impression that the silica content may be slightly enhanced in relation to that of the host rock? There is just a hint of silica increase in some areas closer to the vein although this is debatable.

CHANGEOVER OBSERVATION (TRANSITIONAL ZONE)

This important parameter involves a careful visual inspection to determine precisely what happens to each visible host mineral species as the alteration begins to take effect. Attention is thus focussed on the transition zone. In this instance it is clear that the feldspars (pink-white) have been altered and are seen in all stages of transition from partially to fully altered. It should also be noted that the process also destroys the relatively sharp boundary features of individual feldspars. This edge blurring is a major indication that a mineral is being subjected to alteration. The biotite (dark) remains relatively untouched although very careful observation of some grains give suspicions of edge blurring effects. Although not visible it has probably been partially altered to chlorite? The quartz (grey-white) is untouched.

TEXTURAL OBSERVATION

The partial alteration concentrated within the host feldspars naturally produces a textural result which has a similar pattern to that of the host. It is thus texturally retentive.

CHANNELWAY IDENTIFICATION

The white silica vein contains open void spaces and quartz crystals are discernible at several points. It is thus identified as infill and represents an original channelway. The symmetrical relationship between the vein and the alteration suggests (but does not prove) that the fluids responsible for alteration passed through the channelway now occupied by the quartz infill. This suggests (but does not prove) that the infill and alteration are related. (Obviously many more field examples of the same relationship would be needed to counter the possibility that the silica veining fortuitously formed in this central location at some later date).

It is worth noting that the open space character of the vein suggests a tensional origin for the portion under observation. This could be achieved by any permutation of faulting, joint development, dissolution, or hydraulic jacking. The comb style quartz argues against incremental opening (as opposed to fibre style silica).

CHEMICAL AWARENESS

The major change observable is that feldspars were converted to sericite. The feldspars appear to be predominately potassium rich (K-spars) and the sericite can be tentatively called muscovite (hydrous potassium-rich mica). If this is the case, then firstly water has been added to the rock. Secondly the potassium feldspar with a K:Al:Si ratio of 1:1:3 has converted to muscovite with a K:Al:Si ratio of 1:3:3. This means that K^+ and K_2O may have been lost from the rock. Obviously microscopic confirmation is required to take this subject any further. If the quartz vein is related to the alteration, the fluids also contributed silica at some stage.

PLATE 1A **SERICITIC ALTERATION**
Quartz vein with dark (sericite-dominated) alteration halo in fine-grained granite host.
New St. Patrick copper mine, Copper Firing Line, Herberton, Queensland, Australia.
This plate is presented to allow discussion of the basic principles of alteration (See Section 2 – text) and to give the reader a visual impression of the very common sericitic alteration type.

PLATES 2A, 2B, 2C

GREISEN (SERICITE-SILICA ALTERATION)
Ollera Creek tungsten mine, Paluma District, Queensland, Australia. (Specimen provided by G. W. Clarke).

This specimen has been selected to show an advancing alteration front of diverse mineralogy (silica-sericite) as an example of greisen style alteration, and to demonstrate the texturally destructive nature of silica alteration. The deposit is of the tungsten-pipe style and is a smaller version of the better known Bamford Hill type which is also in north Queensland.

GENERAL ALTERATION RECOGNITION

Alteration is suspected from the readily apparent mineralogical changes as the mineralised (silica-rich) region is approached from the host granite. (*Plate 2A*)

OBSERVATIONAL POSITIONING

Eyeball movement is sufficient with this specimen, although in the field the alteration zone can require physical movement around the 0.5 metre scale.

MINERAL IDENTIFICATION

The alteration zone is a classic example designed to confuse the beginner. Fairly obviously the mineralogy varies and three very vaguely defined zones are apparent (Plate *2A*). An inner grey-white glassy zone and an outer darkish zone are separated by something which is texturally between the two. Having noted this it is clear that all the zones are composed of various combinations of two minerals. Silica (grey-white) predominates towards the right hand edge.

The same silica is present throughout the rest of the alteration zone but is also accompanied by another form of dark grey silica. The second mineral is a dark to pale white mica which is dark coloured on most of the slab surface, but looks much paler on the normal rock face. The mica also has two major forms, occurring as discrete dark coarse crystal clusters within the high-silica (grey-white) zone and as finer-grained slightly paler aggregations within the dark-grey silica zones of the outer regions approaching the granite host. Two other minerals are present in relatively small amounts. Firstly there are some pale pink spots (*Plate 2A – top right, Plate 2C – top right*) and just adjacent to the granite-greisen transition these are some very dark crystal clusters. The pale pink mineral is probably K-feldspar and the dark clusters are probably dark coloured muscovite zones relating to the host rock. (*See below*).

Should this rock be called sericitic alteration, silica alteration, sericite-silica alteration or silica-sericite alteration? Given that the white mica component is mostly rather coarse grained, it could equally well be argued that it be called muscovite alteration. This discussion becomes even more distressing when it is realised that the white micas in the rock are almost certainly unusual and given their environment are probably lithium- or fluorine-rich. Practising tin-tungsten geologists add further confusion by using the very well-established term greisen alteration to describe the coarse -grained mica-silica assemblege. The author would prefer to establish the general eyeball mineralogy and name accordingly and would use the quartz-muscovite prefixes initially, probably acknowledging the traditional greisen term by use of brackets. As can be seen there is no accepted standardisation and at the moment it is just one of those illogical things designed to send the clinically minded into deep depression!

CHANGEOVER OBSERVATIONS – TRANSITIONS

The rock actually contains three transitions (left to right)

(a) Granite to (muscovite) sericite-silica (sharp) texturally retentive alteration
(b) Sericite-silica – to silica-sericite (very vague) - vague texture retentive alteration
(c) Silica-sericite to silica (vague) – new texture (infill)

The granite host rock (*2A, 2B*) is composed of variably sized quartz (white-grey) crystals which exhibit a vague, rounded crystal form, and are commonly linked in chain-like clusters.

These are intergrown with brick red (hematite dusted?) K-feldspars which vary in grain size and range from semi-crystalline to interstitial shapes. Plagioclase is rare but one altered example (sericitised) is present which appears encased by K-feldspar (rapakivi texture) and includes some dark spots. The relatively restricted, interstitial-style dark aggregates are a variety of muscovite. The changeover from pink to grey is quite sharp, and although it is obvious on the broad scale (*Plate 2A*) that the feldspars have been changed this is quite difficult to confirm close-up (*Plate 2B*). The feldspar position is taken up by various forms of dark and pale greisen minerals (micas). There is a tendency for white micas to appear dark on slabs. Some feldspar remnants can be seen in the paler zones (top centre). The granite quartz and dark muscovite seem relatively intact and there are

PLATES 2 A B C **GREISEN (SERICITE-SILICA ALTERATION)**
Ollera Creek tungsten mine, Paluma District, Queensland, Australia. Specimens provided by G. W. Clarke.

hints that extra silica has appeared in the greenish-darkish zones.

Plate 2C depicts the changeover from silica (infill–top right) to silica-rich alteration (bottom left). Compared to *Plate 2B*, it can be surmised that the original feldspar textures are even less obvious, the micas are coarser-grained and there is a general silica increase. This texture has resulted from the continuing alteration of material which started life as that portrayed in *Plate 2B* – that is, an advancing alteration front.

PLATE 2A

PLATE 2B

TEXTURAL IDENTIFICATION

The host rock granitic texture is vaguely retained over the bulk of the zone due to the resistance of the granite quartz to major change. The textural retention breaks down progressively approaching the high-silica edge zone. This rock contains a major trap for both beginner and professional and requires very careful inspection at the silica-rich end with a good knowledge of infill textures (*Plate 2C – top right*). The mica bunches become isolated with strong hints of triangular textures, the silica becomes relatively uniform coloured with hints of crystal faces. Many of the latter can be seen projecting into the mica bunches. The pink feldspar? also shows several hints of triangular interstitial texture. It is very probable that the edge zone conceals a region of infill, with silica texturally-destructive edge-blurring occurring towards the host rock direction (see changeover).

CHANNELWAY IDENTIFICATION

From the infill evidence discussed above it seems reasonable to surmise that a fluid filled space was originally present towards the edge of the specimen. In the field these infill zones form clusters of irregular pipe-like zones and the creation of open-space in the relatively unbrecciated rocks remains a mystery (chemical dissolution is suspected).

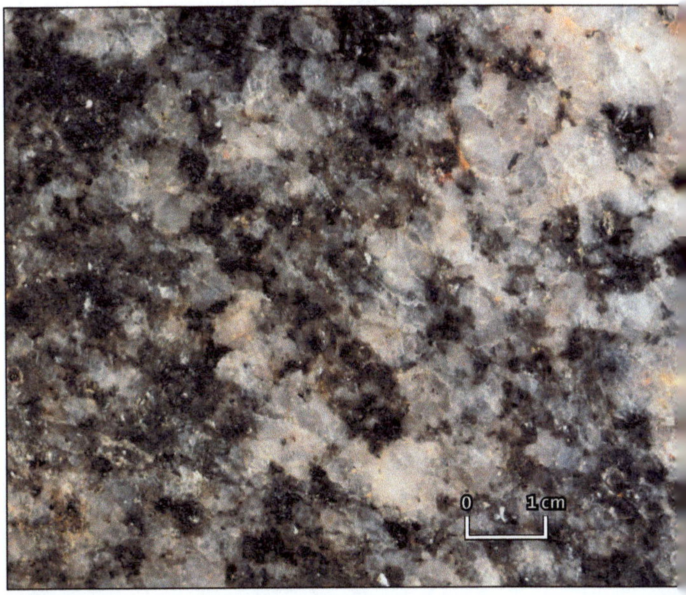

PLATE 2C

CHEMICAL AWARENESS

At the observational scale it can be noted that the silica content would obviously increase from left to right with potassium decreasing as the sericite component diminishes. The change from granite to sericite-silica could result in potassium loss as K-feldspar converts to sericite-silica and any sodium and calcium content represented by placioiclase would be lost.

PLATE 3

SERICITE-SILICA ALTERATION
Phyllic alteration
Chuquicamata copper mine Chile.
(Specimen provided by S. Beams).

This plate has been included specifically to illustrate the porphyry copper alteration style termed phyllic alteration which is commonly associated with stockwork veining.

ALTERATION RECOGNITION

Alteration is suspected by the presence of obvious veins with dark halos.

Observational positioning

Eyeball movement, with close up hand lens inspection.

MINERAL IDENTIFICATION

The alteration zones consist of pale grey-green-white material together with darker minerals which are fine-grained and difficult to identify. The paler minerals are comprised of a grey glassy mineral (quartz) and a paler greenish white component (sericite). The distinctions and colours are best seen associated with the larger vein (top left). This contains a central zone of dark minerals (sulphides?) and quartz which is probably infill. Alteration adjacent to the infill is silica-rich with increasing sericite component approaching the wall rocks. In most circumstances this would be called either sericitic or silica-sericite alteration. However, where found within a porphyry copper system geologists tend to revert to well-established alteration-zoning terminology and call this phyllic alteration (No wonder beginners have problems!) The phyllic alteration zone is particularly important as the vein stockwork is usually the main ore zone. Chalcopyrite and pyrite are the main sulphides. They are not distinguishable on the plate and secondary chalcocite enrichment is suspected as being responsible for the dark colour. It is also noted that dark sulphide spots occur within the wall rocks where they represent sulphide alteration. The full alteration assemblage is silica-sericite-sulphide (pyrite and chalcopyrite).

CHANGEOVER OBSERVATION (TRANSITIONAL ZONE)

The wall rock is a porphyritic igneous rock (monzonite?) with phenocrysts of quartz (dark), plagioclase (white), with rare pink K-feldspar. The latter pink colour is unusual and raises the possibility of pre-existing potassic alteration. The matrix is crystal size (seriate textured) and composed of white/yellowish feldspars and darker quartz. The fine grain size and narrow selvedge zones dictate hand lens scale examination. The principal change is the conversion of matrix feldspars to the pale greenish-white sericite (best seen in the top left vein). As the vein centre is approached (infill zone) the greenish white sericite begins to blur and leaves the impression that silica is increasing. The alteration is probably zoned with silica increasing towards the central silica infill channel (see *Plate 2A*). The smaller veins (bottom right) have little infill and alteration is consequently sericite-dominated.

CHANNELWAY OBSERVATION

Fluid access is obviously via the stockwork fracture system. The main vein channel seems to have been open enough to receive infill precipitation. The smaller veins seem to have been just small cracks which allowed fluid passage but left little or no space for infill precipitation.

TEXTURAL OBSERVATION

Although a little difficult to see, the alteration is texturally retentive at the outer sericite fringes. However, if alteration proceeds to the silica end of the spectrum the inner silica zone is texturally destructive. (NB. Most siliceous alteration is texturally destructive, see *Plate 6*.)

CHEMICAL AWARENESS

(See *Plate 2*)

PLATE 3 **SERICITE-SILICA ALTERATION**
Phyllic alteration, Chuquicamata copper mine, Chile. (Specimen provided by S. Beams).

PLATE 4

CHLORITE ALTERATION IN MIAROLITIC GRANITE
Specimen from the disseminated ore zone at Zaaiplaats tin mine, Limpopo, South Africa.
(Specimen provided by P. J. Pollard).

This specimen has been included to show chloritic alteration and the miarolitic style of permeability channel.

ALTERATION RECOGNITION

Alteration is suspected from the rather fuzzy-blurred zones around the edges of sulphide mineralisation.

OBSERVATIONAL POSITIONING

Although only eyeball movement is required, it is essential that the observer be able to clearly distinguish infill from alteration. This distinction (or lack of!) is a neglected observation and frequently results in misinterpretation of deposit origin. In this case the miarolitic cavity has been filled with mineral precipitates and the fluids have also caused alteration around the edges. The observer thus has to accurately locate the cavity edge.

MINERAL IDENTIFICATION

The infill mineralogy is very varied consisting of dark glassy crystals (quartz) dark green clumps (chlorite), dull silver crystals (cassiterite?) and late silver coloured (arsenopyrite). A few pale-yellow/brown minerals are also present (?).
The alteration mineralogy is best initially observed on the left basal region of the infilled cavity and consists of dark green flecks (chlorite) associated with feldspar. The dominance of the dark green favours the name chloritic alteration.

CHANGEOVER OBSERVATION (TRANSITIONAL ZONES)

The host rock is granite composed of potassium feldspar (red) complexly "intergrown" with albite (pink) and quartz (grey-glassy) Some patches are possibly granophyric with fine specs of quartz intergrown with feldspar. The initial alteration is marked by spots and oriented patches of dark chlorite appearing within the red potassium feldspars (top-middle left), this progresses erratically to eventually take over the feldspar position. All stages of alteration can be found by tracing the declining proportions of red feldspar in the crystals under attack. An area projecting into the top portion of the infill cavity, still retains a little feldspar and can be picked out by the small quartz blebs which seem resistant to chlorite alteration.

TEXTURAL OBSERVATION

The chlorite alteration is texturally retentive as it preferentially occupies the feldspar sites leaving the granite quartz relatively untouched. This sequencing is normal for most forms of alteration although with increasing intensity the textural retention is lost as the host silica is finally replaced.

CHANNELWAY IDENTIFICATION

The fluid has presumably gained access via the miarolitic cavity, which is only one of many within the rock. The chloritic alteration only reaches significant proportions around the cavity and this raises a number of difficult questions regarding fluid access. The granite assemblage is probably in the +600 °C range whilst the hydrothermal assemblage is in the >500 °C range.

Was the fluid always present? – does it move through the rock and if so why? – do the feldspars only show cavity focussed chlorite alteration?

CHEMICAL AWARENESS

The conversion of K and Na feldspars to chlorite requires the addition of Fe, Mg and the loss of K, Na. Presumably some (OH) groupings are also added. It is interesting to note that all the elements added as infill are not recorded by the alteration process which provides considerable food for thought for scientists making deductions on the basis of alteration assemblages alone.

PLATE 4 **CHLORITE ALTERATION IN MIAROLITIC GRANITE**
Specimen from the disseminated are zone at Zaaiplaats tin mine, Limpopo, South Africa. (Specimen provided by P. J. Pollard).

PLATE 5

CHLORITE ALTERATION
Jumna tin mine, Irvinebank, Queensland, Australia.

The specimen has been selected to illustrate a typical fine-grained chloritic alteration product.

ALTERATION RECOGNITION

Alteration would be suspected within this rock from the gradual changes noted from bottom to top approaching the dark green zone. The dark green zone is representative of tin-bearing rocks within the mine.

OBSERVATIONAL POSITIONING

Eyeball movement is the only positioning required for this specimen although within the mine the observer may be required to physically move several metres to reach the edge of the green (chlorite) zones.

MINERAL IDENTIFICATION

The suspected alteration zone contains two prominent minerals in the form of ovate red coloured blobs within a sea of fine-grained dark greenish material. The red blobs are hydrothermal garnet and the dark green material is chlorite. Chlorite is a difficult mineral to identify and many people take some time to pick up the greenish tinge. Alteration chlorite is commonly fine-grained and appears black to the naked eye. Inspection of a broken surface with a hand lens is strongly recommended. Chlorite is also very variable in chemical composition and although usually ferromagnesian based it is commonly iron-rich.

Microscopic work will partially resolve problems but full-scale mineralogical comprehension requires microprobe work. Iron-rich minerals are very susceptible to chloritic alteration. The red garnet is not a common alteration product outside of skarn environments, but may occur associated with iron-rich host rocks such as basalt. The alteration zone also contains minor amounts of a fine-grained pale buff-yellow material (siderite?) and is traversed by a few iron-stained fractures. The predominance of the dark-green material would cause most geologists to call this chloritic alteration.

CHANGEOVER OBSERVATIONS

The host rock (bottom) is composed essentially of variably sized quartz fragments and is known locally as quartzite. There are some minor constituents which are too small to identify with the naked eye. The alteration process is well displayed and quite simply the chlorite replaces the silica fragments via fractures, matrix and grain boundary access. All stages of fluid/rock interaction are apparent with quartz grains occurring in all stages of alteration. Smaller fragments seem to be altered more readily than their larger compatriots. This plate will dismiss any reservations people may have about the ability of some hydrothermal solutions to react with silica!

TEXTURAL OBSERVATIONS

The alteration process would clearly class as texturally destructive, although some remnant ghost structures can be seen on the slab in the least altered regions. These are not easily visible in broken rocks in the field situation.

CHANNELWAY OBSERVATIONS

Within the host rock fluid access is via a series of small fractures which have allowed and enhanced access to grain boundary permeability. The major channelway would be assumed to be towards the top in keeping with the enhanced alteration. It is interesting to note that the red garnet predominates in this region and given its euhedral nature would be a prime candidate for infill. Although barely visible on this surface, the garnet area also contains a lot of chlorite which is slightly paler than in other regions. Again infill is a possibility and could be checked by etching. The ores are in fact strong breccia zones where fragments have been comprehensively altered.

CHEMICAL AWARENESS

The country rock is essentially silica whilst the altered product is essentially chlorite. Obviously large additions of Fe, Mg, Al and (OH) groups are required and have arrived via the hydrothermal fluid.

PLATE 5 **CHLORITE ALTERATION**
Jumna tin mine, Irvinebank, Queensland, Australia.

PLATE 6

SILICIC ALTERATION
Isobella (silver, lead, zinc) mine, Herberton, Queensland, Australia.

This plate has been selected to show typical silica alteration, stockwork style fluid access, the texturally destructive nature of siliceous alteration and the blurred edge effects between infill and alteration.

ALTERATION RECOGNITION

The presence of veinlet style/stockwork sulphide mineralisation with obvious vein selvedge zones immediately raises suspicion of alteration.

OBSERVATIONAL POSITIONING

Within the plate only eyeball movement is required to traverse potential alteration zones. However, within the ore zone, areas of silica/sulphides over several metres in thickness are present. At one stage the failure to recognise infill from alteration caused the brecciated fault zone to be misidentified as a volcanogenic massive sulphide style of mineralisation.

MINERAL IDENTIFICATION

The stockwork vein zones are composed of a grey-white glassy mineral (quartz), a variety of dark coloured minerals (sphalerite) and minor paler yellow spots/blebs (pyrite). The predominance of quartz would quickly suggest that siliceous alteration (silicification) was present. Geologists familiar with veins and infill textures would be cautious concerning the sulphides which are quite probably infill. Within this context it is highly probable that some of the silica is also infill. Additional samples would be sought to resolve the situation.

CHANGEOVER OBSERVATIONS (TRANSITIONAL ZONES)

The host rock (bottom-middle right) is a fine- to medium-grained metasediment composed of silica (grey) and feldspar? (pale cream to pale yellow). It is locally termed quartzite. The changeover is best seen (top-middle left) in zones of intense silica-vein development where segments of host rock are situated between veins and occur in all stages of alteration. The pale cream material is obviously being converted to silica. It is very difficult to see any natural break between suspected silica alteration and suspected silica infill. This edge-blurring effect is very common in siliceous alteration. The difficulty in differentiating siliceous alteration from silica precipitation is a constant problem to both hand specimen and microscopic observers. In most cases the hand specimen observer is at a slight advantage as the siliceous alteration may have a slightly different colour to that of the infill which is not visible in thin section. It is possible in this plate to just perceive the ghost outline of a host rock fragment (top left) where the silica (alteration) is faintly darker than that on either side (infill?) of the ghosted fragment. There is a possibility that more than one phase of silica introduction has occurred as there is a hint of a crosscutting silica-sphalerite vein (centre). More specimens would be required to establish this as the evidence here is a little inconclusive.

TEXTURAL OBSERVATION

The degree of textural retention is very small and the alteration quickly removes most features of the original host rock.

CHANNELWAY IDENTIFICATION

The points of fluid entry are marked by high silica (vein) zones and the rock appears to have been shattered (the specimen is actually an edge to a wide zone of fault breccia extending for approximately a hundred metres along strike). The probability of extensive open space infill is high and given the lack of shearing or rock flour the rock has probably suffered some hydraulic jacking with fluid introduction?

CHEMICAL OBSERVATION

If the pale host rock mineral is feldspar than the altered rocks have clearly lost components such as K, Na, Ca and possibly Al whilst gaining silica. Whether or not the sulphide component has been added depends upon whether the observer interprets them as infill precipitation or alteration. The writer would suspect that the majority are infill components.

PLATE 6 **SILICIC ALTERATION**
Isobella (silver, lead, zinc) mine, Herberton, Queensland, Australia.

◻ PLATE 7

CHLORITE, EPIDOTE (CARBONATE) ALTERATION, Propylitic alteration. Mine dump, Ravenswood gold mining district, Queensland, Australia.

This plate has been selected to illustrate the main features of an alteration style commonly termed propylitic. The rock is coarse-grained and the mineralogical changes are easily observed. Most propylitically altered rocks are fine-grained and emanate from rocks which contain abundant dark ferromagnesian minerals (e. g. basalt, dolerite, andesite).

ALTERATION RECOGNITION

Alteration is suspected from the change in colouration moving from bottom to top. Mineral blurring also increases in this direction and the specimen is in fact a halo to a vein above the top of the plate,

OBSERVATIONAL POSITIONING

Eyeball only.

MINERAL IDENTIFICATION

The altered zone consists of a variety of minerals which include a sharply defined yellowish component together with significant amounts of a grey-green-white mineral, a colourless glassy substance, and a preponderance of dark black-brown to dark green mineral grains. The colourless-glassy mineral is quartz, the green-grey white is partially altered feldspar and the yellow grains are epidote, possibly a Ca-rich variety. The dark minerals are ferromagnesian components and represent biotite and hornblende partially altered to chlorite. Chlorite is probably also responsible for the greenish colouration of many of the feldspars. The alteration assemblage chlorite-epidote (+ calcite) is called propylitic and is a very common low temperature alteration. It is particularly visible in ferromagnesian rocks (basalt, dolerite, andesite) as a greenish tinge (chlorite) to the ferromagnesian minerals with occasional green-yellow epidote.(See *Plate 8*). The name is frequently used in porphyry copper systems where propylitic alteration forms the outer and most extensive zone of a complex zoned alteration system.

CHANGEOVER OBSERVATION (TRANSITIONAL ZONES)

The changeover transition zone is gradual with the intensity of alteration vaguely increasing towards the top. The initial rock (bottom) is a medium-grained tonalite composed of minor quartz (clear glassy), abundant plagioclase (white), and equally abundant dark minerals (biotite, amphibole). Many of the dark minerals contain small yellow spots of fine-grained pyrite. The obvious igneous texture is characterised by sharp grain boundaries.

Although alteration is texturally retentive a close examination reveals that mineral boundaries start to look fuzzy, and the initial sharp interlocking texture becomes blurred. This is primarily achieved at the expense of the plagioclase and dark mineral grains. Good observation entails taking each individual mineral of the original rock and specifically looking to see any changeovers as alteration increases. The quartz grains remain unchanged. The plagioclase grains are variably affected but significant numbers or portions of them are grey-green in colour. It is also apparent that the "yellow" epidote preferentially appears in plagioclase sites. It is probable that it is actually clinozoisite, which is a Ca-rich end-member. Clinozoisite tends to be on the yellow side of the green-yellow "epidote" range. The ferromagnesian minerals (dark) appear to be less altered but careful inspection shows that many of them are not as black or glossy as the originals and have taken on a flat dark green-grey look. This is probably due to chlorite alteration, which is also responsible for obscuring many of the originally sharp grain boundaries. The pyrite spots disappear as alteration increases. Calcite is not really visible at this scale but some of very small bright white spots would come under suspicion?

TEXTURAL OBSERVATION

The alteration is texturally retentive.

CHANNELWAY IDENTIFICATION

The bulk of fluid access has occurred via a millimetre-scale vein channel not visible within the plate. However, careful observation reveals a small vague vertical pale coloured channel (centre right) traversing the altered zone. There is just a hint that alteration is a little more intense at some points adjacent to the channel. The nature of the channel is a little obscure but it seems to be a simple fracture with infill (silica? + chlorite?).

CHEMICAL AWARENESS

Chlorite, although variable in composition, is basically an Mg-Fe aluminosilicate and epidote, although similarly variable, is a Ca-Fe aluminosilicate. The epidote replaces calcium-rich plagioclase as does a

PLATE 7 **CHLORITE, EPIDOTE (CARBONATE) ALTERATION**
Propylitic alteration.
Mine dump, Ravenswood gold mining district, Queensland, Australia.

proportion of the chlorite. It would thus seem that a little Fe-Mg is all that is required to produce the plagioclase alteration. The igneous iron and magnesium-rich minerals have also been partially converted to iron-rich chlorite. It seems probable that much of the alteration is simply a readjustment of elements within the original rock with little addition or loss. Indeed most analytical investigations of propylitic style alteration demonstrate that little has been added other than extra (OH) groupings.

PLATE 8

SILICA, CHLORITE, EPIDOTE, SULPHIDE ALTERATION, Propylitic/siliceous alteration, Intrusive breccia. Esis porphyry copper prospect, Papua New Guinea.

This specimen has been selected to illustrate both the general principles and difficulty of coping with alteration within complex breccia systems. It also provides a second example of propylitic style alteration.

ALTERATION RECOGNITION

The specimen is clearly a fragmental rock. The fragments range from several centimetres in diameter down to dust size particles. There is in fact a complete gradation but without entering into semantic arguments most geologists would see this as a combination of fragments and rock flour. The fragments are generally well-rounded and predominately derived from darkish plutonic-textured igneous rock. There is a hint of quartz fragments (vein material – top left edge). From field relationships together with the above, it has been identified as intrusive (milled) breccia.

Alteration in this instance is suspected firstly by the very fuzzy blurred appearance of the minerals within the coarse-grained igneous fragments, and secondly by the less obvious observation that the smaller size matrix particles are similarly blurred and difficult to discern. The rock also contains a suspiciously high sulphide content suggesting a possible hydrothermal contribution.

OBSERVATIONAL POSITIONING

Alteration within brecciated rocks usually proceeds via matrix permeability and this certainly seems to be the case here. To fully ascertain subtleties the observer should move totally in an attempt to find breccia with unaltered or less altered matrix (that is to seek out the original matrix). Most significantly altered intrusive breccias are a result of late fluids permeating through an originally weakly altered matrix (see changeover observations and timing).

MINERAL IDENTIFICATION

The mineral composition of the rock (on bottom right) is difficult to ascertain in any normal sense. At least two (probably three) fragment types are clearly visible and presumably the milling has resulted in an inhomogenous matrix mix. The large igneous textured fragments are a mixture of quartz, grey-white feldspar, and dark ferromagnesian minerals. The latter are mostly ill-defined and have a dull dark greenish tinge suggestive of chlorite. Many of the white feldspars have equally vague outlines and seem to be replaced by grey silica? A variety of spots (sphalerite, iron carbonate, rutile, titanomagnetite?) occur within the ferromagnesian zones and several patches of green-yellow (epidote?) are present. Some of them seem to be within white plagioclase zones.

The matrix is more difficult to characterise but has a general greenish-grey appearance and in hand specimen is very hard suggesting that it is very siliceous + chlorite. Obviously the dust size particles are derived from the fragments and in this context partially reflect the above description. However, there is a noticeable reduction in white plagioclase material.

From the hand specimen it seems that the main alteration minerals are silica, chlorite, epidote + an unknown pink-brown mineral and minor sulphides. For those who like names, the high silica would be overlooked and the tag propylitic readily applied.

The fragments also contain significant amounts of fine-grained sulphides which are rather difficult to pick out in the plate. These are in the same size range as the prominent pale-pink mineral (titanomagnetite-Ieucoxene?) and at least two species are found. One has a pale yellow tinge (pyrite) and the other is dark coloured with a very dull silver sheen (galena, arsenopyrite?). They are well-represented in the centre-top right fragments and show a preference to appear inside the ferromagnesian mineral component. The matrix contains a similar range of sulphides which are even finer-grained.

TEXTURAL OBSERVATION

Even without unaltered host rocks it is apparent that the bulk of the altered fragments have a high degree of textural retention and have been derived from a medium-grained igneous rock composed originally of feldspar, quartz, and ferromagnesian minerals. The high percentage of ferromagnesian minerals suggests something in the tonalite-monzonite range. The matrix is reasonably texture retentive despite the silicification. The grit scale particles are still visible although many of the smaller ones are "ghosted" or blurred by silicification.

CHANNELWAY IDENTIFICATION

The most significant alteration effects are within the lowest particle size range of the matrix material, and this would be interpreted as indicating that fluid access was via permeation of a relatively unconsolidated gritty/dusty breccia matrix, with an associated secondary access to the fragments.

PLATE 8 **SILICA, CHLORITE, EPIDOTE, SULPHIDE ALTERATION**
Propylitic/siliceous alteration, intrusive breccia.
Esis porphyry copper prospect, Papua New Guinea.

CHANGEOVER OBSERVATION AND TIMING PROBLEMS

At first sight the pervasive alteration style suggests that a fluid has simply permeated through the breccia matrix. The fluid could either come as part of the breccia formation or more likely arrive at some later date. This problem could only be resolved by applying the first principle of alteration – MOVE – to ascertain if there are areas of matrix remaining in their pristine state. An experienced observer would also have reservations concerning the high degree of silicification and sulphide presence linked to the chlorite/epidote (propylitic) association – possibly even pre-breccia. It is possible that the propylitic stage is in fact over printed by a later silica-sulphide alteration. This question might be resolved from well-directed microscope work but once again the best solution would be to MOVE, and inspect the less silicified zones of the breccia, and also the unbrecciated host/source rocks. This rock is not really for the beginner, and could induce a misleading, false sense of propylitic security.

CHEMICAL AWARENESS

The timing comments render full chemical appraisal difficult, and it should be noted that given this and the mixed fragment nature, a chemical analysis is not wildly helpful. The rock has obviously gained in silica, and some sulphide components, and judging from the state of the plagioclase feldspars has lost some of its components (mostly calcium). The iron and magnesium component represented by chlorite/epidote may only represent readjustment of pre-existing ferromagnesian minerals (see propylitic discussion – *Plate 7*).

◻ PLATE 9

POTASSIUM FELDSPAR (K-FELDSPAR) ALTERATION Potassic alteration. Chuqui Norte Chuquicamata copper mine, Chile.

This specimen has been selected to illustrate one style of potassic alteration within a porphyry copper system.

ALTERATION RECOGNITION

Alteration suspicions are aroused by the presence of a stockwork of numerous small fractures associated with pink vein selvages.

OBSERVATIONAL POSITIONING

Only eyeball movement is required. Observers should however, be aware that most potassic alteration in porphyry systems is at the micro fracture level and may not be not immediately obvious. Even within seemingly glaringly obvious examples the pale pink colouration is easily overlooked in an unslabbed hand specimen. This emphasises that all alteration observation should be quickly supported by hand lens inspection.

MINERAL IDENTIFICATION

The most visible feature is the pale pink colouration associated with the stockwork system. The veins are composed of a grey vitreous mineral (quartz) and also sporadically contain small dark minerals (sulphides? biotite?). Pink alteration systems may prove difficult to identify by eye, due to the wide range of potentially pink alteration products. (K-feldspars, albite, hematite, carbonate, silica). However, within the porphyry copper context potassium feldspar is a good guess and porphyry geologists would have little hesitation in nominating potassic alteration. Once again conventional usage creates immense problems for beginners as the term potassic alteration gives only vague clues concerning the nature of the specimen.

The two common potassium-rich minerals which characterise this style are K-feldspar and biotite. Unfortunately they occur in extremely variable proportions and in a wide variety of structural styles. Sometimes the K-feldspar component is dominant but more commonly the biotite dominates. This is especially true within the more mafic porphyritic hosts which predominate within the Papua New Guinea-island arc porphyry systems. Even within K-feldspar dominated systems recognition is not assisted by the common presence of white K-feldspar! Generally speaking potassic alteration is difficult to detect as the rocks look fresh with biotite-feldspar alteration mineralogy looking very similar to the biotite-feldspar of the host rocks. The structural styles range from subtle stockworks in porphyries, through to very subtle semi-pervasive alteration of fine-grained breccia matrices or grain boundary permeation through fine-grained pyroclastic, volcaniclastic and sedimentary rocks.

The observer is requested to take note of some of the very fine impersistent cracks in the top left quadrant. The dark component in these is quite probably biotite. This dark small-scale crack style is used by experienced observers to pick up potential potassic alteration and is actually more diagnostic and useful than the pink effect which dominates this particular specimen.

CHANGEOVER OBSERVATIONS (TRANSITIONAL ZONES)

The pink alteration is not uniformly distributed along the small fractures and a close inspection reveals that it shows a distinct preference for the feldspars of the fine-grained matrix material whilst leaving the quartz unaffected. Similarly, small feldspar (white) phenocrysts are more prone to alteration than their larger counterparts.

The proportion of small dark spots (sulphides?) exhibits a close (although not perfect) spatial relationship with the alteration. At the very subtle level an experienced observer might suspect that the ferromagnesian component of the monzogranitic porphyry host rock (biotite? hornblende?) is looking a little fuzzy/blurred around the edges. That is to say the dark grain boundaries are not sharp as they are in normal igneous rocks, and in places contain hints of a pale brownish colour. It is highly probable that these have been partially converted to secondary biotite.

TEXTURAL OBSERVATION

The alteration is texturally retentive with most features of the original rock remaining visible.

CHANNELWAY IDENTIFICATION

The fluid access is obviously via an extensive stockwork fracture system which is present at all scales. This incredible shattering is a characteristic feature of most porphyry copper systems and only rarely reaches this extent in most other mineralisation styles. Careful observation shows that the fluid prefers to gain secondary penetration via the finer-grained matrix component.

CHEMICAL AWARENESS

The chemical changes relating to alteration obviously depend on the composition of the crowded feldspar phenocryst component, the matrix feldspars, and whether or not the ferromagnesian minerals have

PLATE 9 **POTASSIUM FELDSPAR (K-SPAR) ALTERATION**
Potassic alteration,
Chuqui Norte, Chuquicamata copper mine, Chile.

been altered to secondary biotite. The hint would be that there is a small potassium increase accompanied by loss of sodium and calcium (from the feldspars). If alteration of the ferromagnesian minerals has occurred there could be some iron or magnesium loss. However, secondary biotite is commonly magnesium-rich. The sulphide component with the exception of iron would be introduced.

■ PLATE 10

ADULARIA ALTERATION
Andesitic volcanics.
Cracow gold mine, Queensland, Australia.

This plate has been selected to illustrate adularia, pervasive alteration, and fragment destruction due to fluid access via matrix and fracture networks.

ALTERATION RECOGNITION

The main features which suggest the presence of alteration are the rather blurred boundaries to some of the more obvious fragments (central zone) and the suspicion that there are other fragments which have been "ghosted out" (top left). These features when linked to the vague hint of veining (bottom right) are sufficient to suspect extensive modification of a fragmented rock by hydrothermal alteration.

OBSERVATIONAL POSITIONING

The suspected alteration style is fairly pervasive and physical movement of the observer would be required to locate potential host rocks. In reality these are mildly sericitised green-grey fragmental andesites.

MINERAL IDENTIFICATION

The most visible components are orange-red/pink feldspars most of which occur as phenocrysts within fragments. The feldspars are set in a fine-grained green-grey (silica, sericite?) matrix. The vague vein-like areas similarly contain orange-red/pink feldspar together with yellowish pale-orange material (silica). The remaining texturally diffuse areas are composed of similar combinations of the above minerals. Some small dark minerals are also evident (iron oxides?). In this instance none of the minerals are readily identifiable and without visible host rocks it is difficult to nominate those which are the result of alteration. Given a host rock it would be evident that white feldspars have been converted to pink and that there has been considerable introduction of silica. As previously indicated (***Plate 9***) pink alteration is always difficult to identify with certainty and could represent potassium feldspars, albite, silica, hematite or carbonate. In this instance X-ray Diffraction and microscopy have established that the new feldspar is adularia and belongs to an epithermal alteration assemblage.

CHANGEOVER OBSERVATIONS

The suspicion of extensive alteration would naturally lead the observer to check laterally for less-affected host rocks. No coherent observation can be made from this plate.

TEXTURAL OBSERVATION

The alteration is reasonably texturally retentive in that the fragments are recognisable as porphyritic igneous rocks. Details of fragment boundaries and matrix textures are however a little vague and it is difficult to speculate on breccia style.

CHANNELWAY IDENTIFICATION

Judging from the distribution of intensity of alteration the fluid access seems to have been via the matrix component of the breccia. Although the fragmental-breccia style is uncertain there are hints of infill zones given by the two paler-yellowish areas in the centre of the plate. A thin curved strip and a more arcuate zone rimmed by red feldspar could represent direct precipitation into a fluid-filled gap. The observer should obviously move to locate clearer examples, but it looks like the fluid pathway will be via open channels (breccia) and matrix permeability.

CHEMICAL AWARENESS

Without more information concerning host rocks and mineral identification it is not possible to speculate too far upon possible element mobility. A silica increase seems a strong possibility, and an introduction of potassium.

PLATE 10 **ADULARIA ALTERATION**
Andesitic volcanics,
Cracow gold mine, Queensland, Australia.

PLATE 11

BIOTITE ALTERATION, Potassic alteration. Main Pipe Breccia, Mt Leyshon gold mine, Queensland, Australia.

This plate has been selected to show biotite alteration and breccia style fluid access. It also illustrates some of the complexities involved with breccia alteration.

ALTERATION RECOGNITION

Alteration is difficult to detect and the specimen at first sight could well represent a normal clastic sedimentary rock. Some suspicion might be aroused by the fact that clasts are difficult to recognise below about grit size. The specimen is actually an intrusive (milled) breccia with a strongly altered matrix.

OBSERVATION POSITIONING

The key to positioning hinges upon the observers experience with breccia systems. If the possibility of either intrusive style breccia or matrix alteration is recognised, the procedure is obviously to move and examine the situation within other parts of the system. In most intrusive breccia systems major alteration occurs subsequent to breccia formation and it is possible to move from areas of relatively unaltered breccia matrix to zones of strong alteration associated with late hydrothermal fluid introduction (See channelway section below).

MINERAL IDENTIFICATION

If alteration is suspected it would be represented by the fine-grained dark material of the matrix. The dark colour could be caused by a variety of minerals such as tourmaline, silica, chlorite or biotite. In this case tourmaline and silica would be eliminated by hardness testing. The black, rather than black-green colour, would suggest biotite but a thin section would be required for confirmation.

CHANGEOVER OBSERVATION (TRANSITIONAL ZONES)

Changeover observations within intrusive breccias systems require careful consideration. This is particularly true where matrix alteration is intense. In such situations although alteration of the matrix is relatively uniform, the response of the different fragments is variable. The amount of fragment alteration ranges from complete to very little, and there is no guarantee that alteration response will be mineralogically similar for differing fragment compositions. The problem is compounded by the following.

1. The fragments may have become altered as a result of hydrothermal fluid interaction before the brecciation process.
2. The fragments may have been altered as a result of hydrothermal interaction as part of the breccia forming process. This would affect both pristine and previously altered host rocks.
3. All of the above may have occurred and a new major alteration may occur as a result of late-stage fluids permeating the breccia. This situation is in fact the norm with mineralised intrusive breccia systems. This will affect both previously altered fragments and may extend to react with relatively fresh rocks in the margins of the breccia.

The above complications serve to reemphasise the prime alteration rule which can be simply stated as move. An excursion into the breccia margins and wall rocks will usually resolve most of the problems relating to 1) above. An attempt to locate areas of breccia which have not been overprinted by late-stage fluids will similarly help with determining the effects due to breccia emplacement. This process is assisted considerably by utilising drill core. In most situations the problems prove to be less horrendous than might be supposed, but the author would confess that there are a few situations which may require months of diligent work to fully resolve the sequencing.

The fragments within the plate are clearly of variable origin and have been variably affected by the late-stage biotite-dominated alteration. Some are "ghosted" out at the small scales, whilst others appear unaffected. The pink-tinged fragments are mostly of the fine-grained granite/porphyry mode and might possibly have been affected by feldspathic (K-Feldspar) styles of alteration. The yellow tinge within the bottom centre pink fragment is epidote alteration of feldspar. Several mysterious rim style alterations are clearly visible notably on the large grey green fragment and also around the pink fragment directly to the right of it. In both cases the rims are not actually continuous and it is very uncertain whether or not the rims relate to alteration caused by the late biotite event or to a previous event during breccia formation. They could equally well relate to both, with the late overprinting event reacting differently with the variable results of a previous alteration!! Again this emphasises the requirement to MOVE (preferably with a purpose). The bright yellow colour associated with the small crack is surface oxidation (top).

TEXTURAL OBSERVATION

The degree of textural retention within the fragments is relatively high and most geologists would be suggesting the host rocks mentioned above. The textural retention progressively decreases with fragment size. Smaller fragments of appropriate compositions become blurred around the edges and eventually invisible to the eye. In general terms this process is best seen

PLATE 11 **BIOTITE ALTERATION , Potassic alteration, Main Pipe Breccia, Mt Leyshon gold mine, Queensland, Australia.**

at around "grit" size. In other words, if the fragments are reactive the smaller grains are altered first.

CHANNELWAY IDENTIFICATION

The mode of fluid access is assumed to be via the permeability afforded by a relatively unconsolidated breccia matrix. Given the style of breccia very little open space would be available and visible infill textures rare.

CHEMICAL OBSERVATION

Most of the alteration reactions are obviously complex given the nature of the rock (see changeover section above). However, at the broad scale it is clear that a significant amount of iron and potassium must have been present within the late fluid to account for the biotite. However, without knowing the general chemistry of the "original" matrix it is very difficult to speculate as to the extent of local versus outside derivation. Given general knowledge concerning biotite alteration a significant outside component is suspected, involving a high temperature saline magmatic solution. This is further supported by the apparent stability of feldspar-rich fragments in the matrix. It is essentially potassic alteration similar to that which is related to conventional porphyry copper systems.

PLATE 12

ALBITIC ALTERATION
Mallae Gap Creek, Selwyn Range, Queensland, Australia.
(Specimen provided by P. J. Williams).

This specimen has been included to show an albite alteration and to illustrate the problems concerned with fine-grained alteration products.

ALTERATION RECOGNITION

Alteration is suspected from the nature of the gradational transition from the dark host rock to a pale product. In the field, the alteration forms part of a fracture controlled vein-like structure.

MINERAL RECOGNITION

The suspected alteration product is very fine-grained and contains prominent flakes of a lath-like yellow-pink mineral (?) occurring within finer-grained pale pink material (feldspar?) speckled with darker grey-glassy wavy elongate grains (quartz). Small dark spots are scattered throughout and there are vague areas which seem a little pinker than average (hematite?).

It would be a very confident geologist who could pick this assemblage from hand specimen alone, and petrological support is necessary for confirmation. The general white/pink colouration suggests feldspar but hematite, carbonate, and silica are all valid alternatives. In the field it would be best described by colour/texture awaiting petrographic confirmation. This is commonly the case with fine-grained end products, and albite in particular can assume a wide range of colours (white, pink, red, grey-glassy). For completeness the reader will be interested to learn that the alteration is predominately albite (pale flesh-white), associated with minor sphene (yellow-pink), quartz (dark grey) and magnetite (dark). The above description was written before inspection of a thin section.

CHANGEOVER OBSERVATION (TRANSITIONAL ZONES)

The host rock is a dark schist composed of fine-grained minerals. The main mineral components are dark ferromagnesian minerals (biotite? amphibole?), darkish green areas (chlorite? sericite? altered feldspar?) white-grey-glassy minerals (quartz), and some minute bright white spots (?).

Within the transition zone, the ferromagnesian component disappears, the glassy component remains. The remainder becomes increasingly dominated by the greenish mineral, which grades to the dominant paler pink colour of the main alteration zone. The prominent pink-yellow mineral of the alteration assemblage starts to appear within the green transition zone.

TEXTURAL OBSERVATION

The alteration is very texturally retentive and retains the prominent schistose character of the host.

CHANNELWAY IDENTIFICATION

As depicted it is difficult to assess whether or not the fluid channelway is related to a fracture system beyond the upper plate margin, or whether it is layer controlled. Field information would confirm the vein-fracture style.

CHEMICAL AWARENESS

It would be difficult to comment in this instance without microscopic mineralogical support. However, knowing that albite is the principle new mineral, it seems that considerable Na has been added at the expense of Fe, Mg and possibly silica.

PLATE 12 **ALBITIC ALTERATION**
Mallae Gap Creek, Selwyn Range, Queensland, Australia.
(Specimen provided by P. J. Williams).

PLATE 13

CLAY-CARBONATE ALTERATION
Argillic alteration.
Mine porphyry, Mt Leyshon gold mine, Queensland, Australia.

This specimen has been included to show the effects of argillic alteration. It is however, a complex rock and also shows features of sulphide and carbonate alteration.

ALTERATION RECOGNITION

The rock arouses suspicions of alteration owing to the pervasive white colouration which is linked to the soft clay-like nature of the hand specimen.

OBSERVATIONAL POSITIONING

The white (clay) is pervasively distributed and is another example which would require considerable movement by the observer. This specimen would be a bad place to start, and movement of tens of metres is required to locate less-altered/fresh zones of the porphyry. The picture is also confused by the vein-like sulphide zone (lower-right) which could be an overprinting later phase. This also requires that the observer move to try to define the relationship between sulphides and clay. There is much to resolve concerning paragenesis.

MINERAL IDENTIFICATION

Given the above complications it would be unwise to proceed too far with this rock pending overprinting/paragenetic resolution. However, the minerals present are a fine-grained white background (clay) with phenocryst-style minerals (top left) including a pale buff brown mineral (siderite) a grey mineral (quartz), a dark mineral (sphalerite?), a pale green mineral (sericite?), and an rare small shiny spots (pyrite). With the clay predominance the term argillic would not be inappropriate, although it would be better to quote the assemblage and look for any timing relationships. Similar assemblages occur throughout the rock although it is probable that the coarser-grained sphalerite-pyrite within the vein area is infill.

CHANGEOVER OBSERVATION (TRANSITIONAL ZONES)

Even without any original rock for comparison it is apparent that two different domains are present within the plate. The high phenocryst zone (top left centre) gives a strong impression of having originally been a porphyritic igneous rock, whilst the remainder, although containing isolated scattered phenocrysts, is more enigmatic in terms of origin.

The porphyritic zone contains clear shapes which suggest mineral origins. The quartz blobs look like original quartz, the brownish-dark materials could easily have been feldspar, and the green shapes could easily have been a mica (biotite?). This has in fact been confirmed from other specimens, which also show that the original porphyry is strangely heterogeneous with patches of high crystal concentrations occurring within zones of more sporadic phenocryst/high matrix texture. A very close examination reveals that many of the carbonate spots contain smaller dark and pale sulphide spots (sphalerite/pyrite) and that the intensity of sulphide alteration increases adjacent to the small veins (especially bottom, centre left) where entire phenocrysts have been altered/replaced. This observation opens up a realm of speculation and a case could be made that the carbonate/clay alteration precedes the brecciation and sulphide introduction with the carbonates providing favourable sites for sulphide formation (carbonate replacement!). The alteration paragenesis is obviously complex but does illustrate the need for very careful systematic observation in situations which are not simple veins. This rock is not really for the beginner but will help set the scene for a later volume on paragenesis.

TEXTURAL OBSERVATION

From the above discussion it seems that both the clay and sulphide styles are essentially texturally retentive.

CHANNELWAY IDENTIFICATION

The fluid access examination faces the same problems as with timing above. The sulphide vein areas are clearly part of a brecciating fracture system which has created some open spaces and accompanying fractures. The small-scale breccia system is dominated by infill and there is little or no sign of any increase of alteration intensity adjacent to the sphalerite-pyrite veinlets. This raises suspicions that the more pervasive clay alteration belongs to an earlier (or later) system. The clay alteration is pervasive and shows little sign of fracture control.

CHEMICAL AWARENESS

Serious chemical consideration obviously requires paragenetic resolution, and a good understanding of the original host. Assuming the clay to be kaolin or illite, a major loss of alkalis (Na, K) and calcium could be anticipated. Obviously metallic components (Zn) and sulphur have arrived at some time.

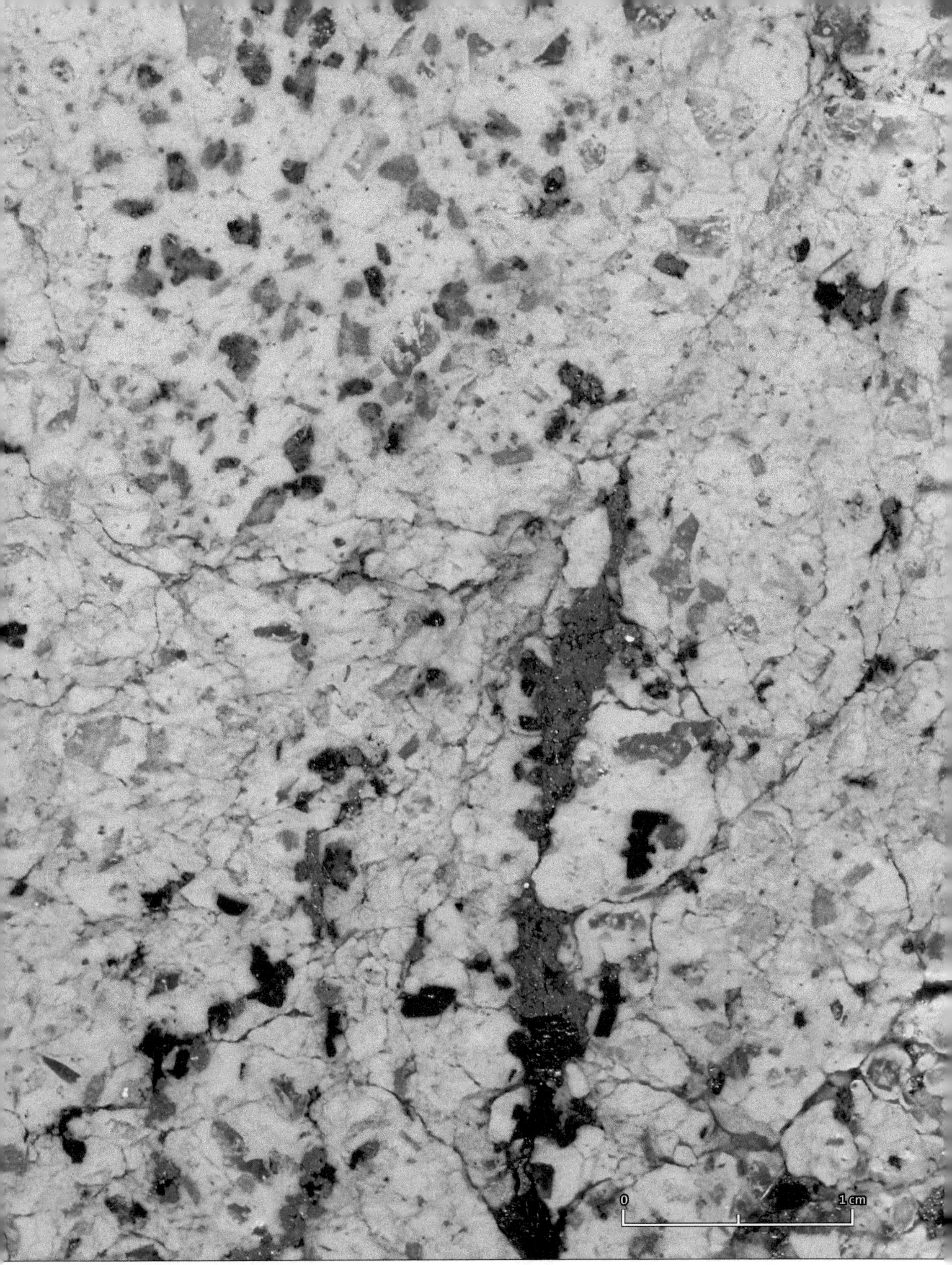

PLATE 13 **CLAY-CARBONATE ALTERATION**
Argillic alteration,
Mine porphyry, Mt Leyshon gold mine, Queensland, Australia.

PLATE 14

SILICA-ALUNITE-PYRITE AND PYROPHYLLITE-ALUNITE-PYRITE ± DIASPORE ± SILICA ALTERATION
Advanced argillic alteration.
Temora gold mine, New South Wales, Australia.
(Specimen provided by A. Allibone).

This specimen has been selected to illustrate the difficulties involved with accurate identification of clay minerals and to show the general style of textures involved with alteration in high-strain zones.

ALTERATION RECOGNITION

Within the frame of the photograph it is rather difficult to pick out potentially altered rock. Some suspicions are aroused by the anastomosing crosscutting ductile style zones (centre left) and the correlating inference that the dark elongate ellipsoidal features represent some form of host rock that has been altered. The actual hand specimen offers further clues in that it is mildly soapy to the touch, which indicates some kind of clay enhancement.

OBSERVATIONAL POSITIONING

This rock is another example of a situation where the observer is best advised to physically move. The key feature is the realisation that the entire rock represents an area of extremely high strain (layer shortening-flattening-foliation, boundinaging-folding). Ductile deformation at this level of strain will naturally reconstitute the original rocks and also form favourable permeable sites for fluid movement. The best chance of approaching original rock compositions lies in locating wall rocks occupying portions of the system where heterogeneous strain distribution has preserved original rocks. These are commonly located towards the margins of the high strain zone and in many cases are represented by the adjacent wall rocks. It will ultimately become important to separate effects due to the original deformation from those formed at later dates. Reactivation and later stage fluid introduction are the norm. Resolving overprinting features is often a difficult task!

MINERAL RECOGNITION

Given the fine-grained nature of the assemblages accurate hand specimens identification is impossible. This is true of practically all styles of clay-dominated alteration products and ultimately specialised equipment such as a PIMA is required for full scale mineral identification. Geologists with limited experience of clay-dominated alteration styles should not be disappointed that they are unable to distinguish the major clay varieties. This particular specimen has been well-studied and is known to contain two major assemblages, silica-alunite ± pyrite-gold and pyrophyllite-alunite-pyrite ± diaspore. Both forms of alteration are conventionally termed advanced argillic. The alunite-silica assemblage predates the pyrophyllite-silica and is represented by darker regions occupying the bottom half of the plate.

It is possible that some of the white flecks in the dark zone are alunite?

The paler zones contain the bulk of the pyrophyllite which is also probably responsible for the background pale greenish hue throughout.

CHANGEOVER OBSERVATION (TRANSITIONAL ZONES)

Without information concerning the host rocks and alteration mineralogy it would be difficult to discern much from this specimen. The greenish coloured material certainly appears to take over the grey matrix of the darker rocks in several places. This is best noted around the elongate edges of the darker lenses in the central zone. Again the fine-grained nature of clay alteration renders hand lens observation a little difficult. The host rocks are actually Ordovician volcanics and minor volcanic lithic sediments.

CHANNELWAY IDENTIFICATION

The later pyrophyllite-rich zones overprint the darker alunite-rich zones and judging from the greenish concentrations associated with the crosscutting high strain (shear) system (centre region) are a result of fluid access during or subsequent to a period of reactivation located within the "shear zone". Fluid access is presumably controlled by higher permeability within this region. The controls on the earlier alunite-rich system are probably similar although not obviously discernible from this sample. Gold and minor pyrite are associated with the alunite-silica assemblage and the rock is part of the main ore zone at Temora. As would be anticipated infill textures are lacking.

TEXTURAL OBSERVATION

It is fairly obvious from the rock that the alteration is texturally retentive with a variety of deformation fabrics remaining well-preserved. These include high strain induced foliations, and flattened elongate boundinaged "wall rock" fragments.

PLATE 14 **SILICA-ALUNITE-PYRITE AND PYROPHYLLITE-ALUNITE-PYRITE ± DIASPORE ± SILICA ALTERATION, Advanced argillic alteration, Temora gold mine, New South Wales, Australia (Specimen provided by A. Allibone).**

CHEMICAL AWARENESS

From the hand specimen alone there is little that can be noted. With uncertain host rocks and uncertain alteration, observational chemical inferences are precluded. However, it is worth noting that advanced argillic alteration normally results in substantial element adjustment and is sometimes referred to in porphyry systems as alsic alteration. This refers to the dominance of silica and aluminium usually at the expense of Ca, Mg, K, Na and Fe. Pyrophyllite is essentially an aluminium silicate $Al_2\ Si_4\ O_{10}\ (OH)\cdot$ Alunite does in fact contain some of the above and is essentially a hydrated potassium aluminium sulphate $(K, Na)\ Al_3\ (So_4)_2\ (OH)_6$

◘ PLATE 15

HEMATITE ALTERATION
Red rock alteration.
Corella formation, Cloncurry Region, Queensland, Australia.

This specimen has been included to show both layer, and layer selective style alteration.

ALTERATION RECOGNITION

Alteration is suspected in this specimen as the layered gneissic rocks become progressively more reddened towards the right. The alteration is even more obvious in the field as red coloured zones extend out in a Christmas tree/telegraph pole style from fractures traversing the metamorphosed Corella formation.

OBSERVATIONAL POSITIONING

Eyeball movement is all that is required with this specimen, although in the field the alteration can extend outwards for 1-2 metres. Layer replacive alteration is, very common and can extend for tens of metres from vertically traversing fractures. In such cases a short walk is required. If the feeder fractures are subparallel to the layers the alteration can extend up to the 0.5 kilometre scale and a major hike becomes essential. Sulphides are very common alteration minerals, and all layered sulphide deposits require automatic inspection of their extremities to seek potential changeover zones. The MOVE rule is very important for this style.

MINERAL RECOGNITION

The alteration mineralogy within the specimen is obviously dominated by the red mineral. As previously remarked red/pink minerals are common alteration products including K-feldspar, adularia, albite, carbonate, silica, and very commonly hematite, and hematite dusting of feldspar. A prudent move here would be to delay nomenclature pending microscope work.

Given the brick red colour there is a strong possibility that hematite is involved. A minor dark grey reflective component (magnetite) is also present, as are some bright whitish spots (pyrite?) and yellow-green areas (epidote) in the bottom right. These could belong to the host rock assemblage.

CHANGEOVER OBSERVATION

The host rock has suffered amphibolite grade metamorphism and from hand lens inspection is composed of narrow gneissic layers which are mineralogically distinct (top left) with various combinations of quartz (dark grey), feldspar (greenish grey), magnetite (dark reflective grey), ferromagnesian minerals (amphibole? chlorite?) with minor epidote (green) and pyrite (white). The latter two are sporadic in occurrence and could be part of the hydrothermal (or another hydrothermal) assemblage. Alteration occurs principally by conversion of feldspar to the pink mineral and thus predominates in felspathic layers. Assuming that alteration proceeds from right to left it is evident that the fluids have penetrated further along some layers then others. Some of the dark layers (semi-parallel microveins?) are less-affected possibly because they are composed of silica-magnetite.

CHANNELWAY IDENTIFICATION

The main fluid pathways are layer controlled, with the main channel lying somewhere to the right of the photograph. Some indication of the nature of the ingress channelways can be seen from the very faint vertical feeder crack (right hand side) marked as a red zone (top right) which actually traverses the entire frame. The trace of this is closely associated with a pyrite patch (bottom right). In the field the alteration seems to extend a considerable distance from apparently very small feeder fractures. Alteration predominates massively over infill.

CHEMICAL AWARENESS

Identification of the red alteration is required to allow any major chemical inference. Petrographic examination confirms that this is hematite. The host rock contains considerable iron, in both ferromagnesian mineral silicates and in magnetite. It is probable that the iron has been derived from these sources (oxidised) rather than being sourced via the fluid. However, the fluid must have been sufficiently oxidising to convert these other minerals to hematite.

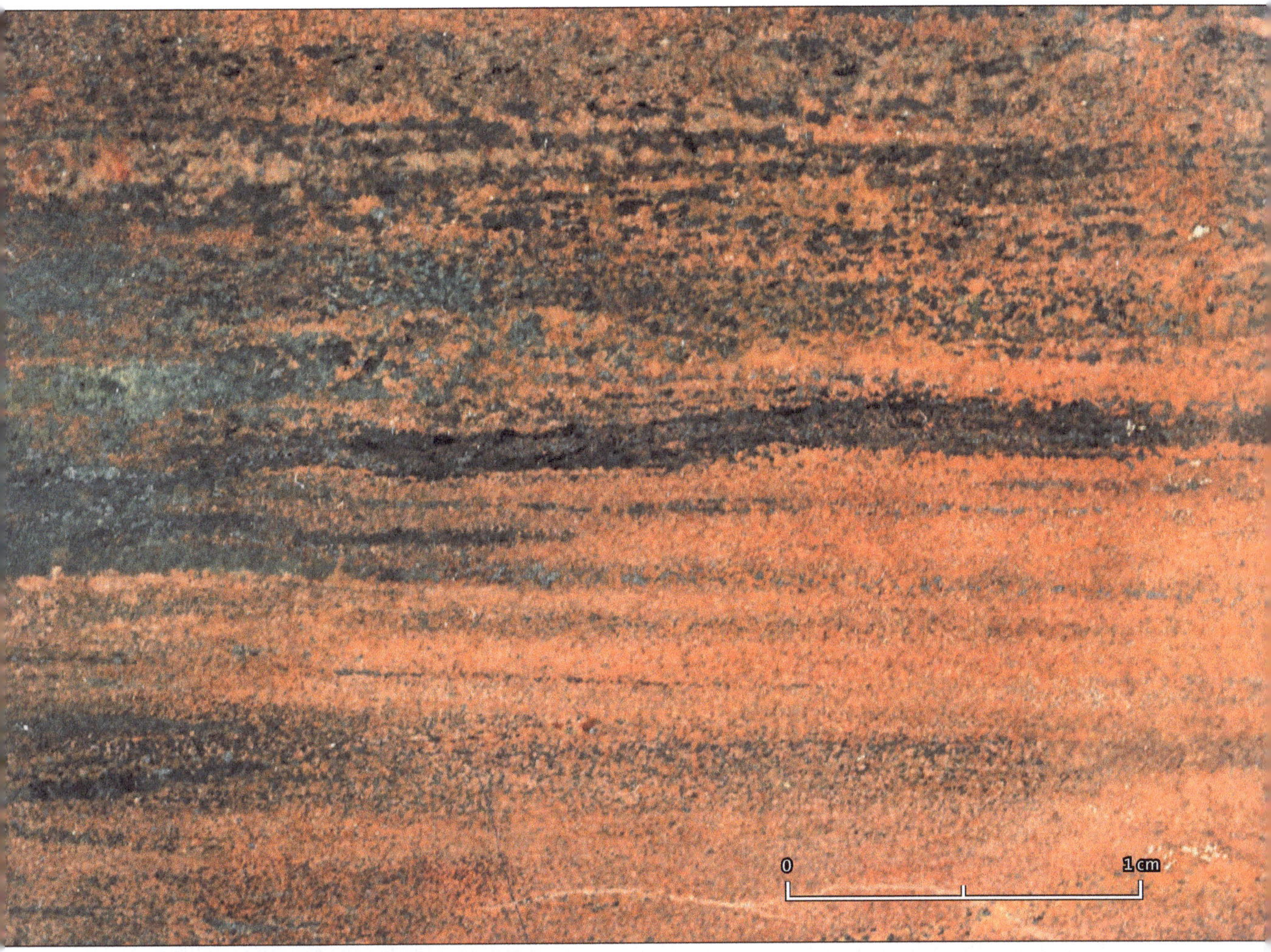

PLATE 15 **HEMATITE ALTERATION, Red rock alteration.**
Corella formation, Cloncurry Region, Queensland, Australia.

◻ PLATE 16

MAGNETITE-CHLORITE-ALBITE ± EPIDOTE ALTERATION.

Leeuwpoort "C" tin mine, Rooiberg, Limpopo, South Africa.

This specimen has been included as an example of a zoned alteration halo, and also to illustrate magnetite as an alteration product. It also serves to illustrate the difficulties in identifying fine-grained alteration products.

ALTERATION RECOGNITION

Alteration is easily suspected as the darker zone occupying the area between vein infilling (top) and wall rock (middle-base).

OBSERVATIONAL RECOGNITION

The only observation positioning required is simple eyeball movement.

MINERAL IDENTIFICATION

The zone of interest is obviously rather variable and at the broad scale comprises a zone of infill (white-dark crystals), succeeded successively by a pale greenish zone, a metallic grey zone, a pale pink-white zone, and a pink zone representing the host rock. The infill zone is composed of carbonate (white) deposited on top of dark crystals (cassiterite-dark brown, magnetite-black-metallic). The pale green layer is composed of pale whitish green lenticular blebs (epidote? sericite? feldspar?) with darker green interstitial component (chlorite?) and rare some dark metallic fine-grained spots or patches (magnetite). The metallic grey zone is composed of metallic grey material (magnetite) and pale pink or green spots, blebs crystals? (feldspar? epidote?). The pale pink zone is transitional to the pink wall rock and is composed of very small dark grains (quartz) and pale pink grains (feldspar?) with minor some vague larger patches which are darker. Mineral identification at this grain size is difficult with only a hand lens, but it is clear that magnetite and probably epidote are major components. In thin section this rock actually contains a great deal of albite which was not recognised in hand specimen. All the minerals tentatively labelled feldspar? above proved to be albite.

CHANGEOVER OBSERVATION (TRANSITIONAL ZONES)

The zoned nature of the alteration assemblages require that each changeover be considered separately.

1. The host rock

The host rock consists of very fine-grained pink feldspar? intergrown with dark quartz of similar grain size. Also present are larger vaguely circular blebs and clusters of darker material (mica?) which ranges from dark to paler green. The paler green could be an alteration product of darker material and the same colour occurs in vague areas possibly representing alteration of the background feldspar. There is a loose correlation between the vaguely altered areas and the paler green large dark clusters. The green colour could be sericite, chlorite or epidote. The pink colours may be due to hematite alteration of potassium feldspar. The host rock is fine-grained granite.

2. Transition to pale zone

The pale outer alteration zone is texturally coherent with the host and seems to develop the pale colour as red feldspars convert to pale-pink to white albite. Minor magnetite is randomly present. The pale coloured feldspar could be potassium feldspar without hematite, albite, or a clay mineral? (Subsequently petrology confirmed albite).

3. Transition of pale zone to grey metallic zone

The grey metallic zone is vaguely texturally coherent with the pale zone and seems to result from magnetite developing at the expense of a considerable amount of feldspar (and /or quartz?)

4. Transition from the grey metallic to the pale green zone

This transition results in loss of textural continuity with the previous zone. Some of the anastomosing darker material which is almost stylolitic in places may represent a strain fabric. The rock is composed of magnetite, chlorite, and a pale mineral with a greenish tinge (albite).

TEXTURAL OBSERVATION

These have been covered above, but this specimen represents an example where textural retention decreases approaching the fluid source. The latter is represented by the infill zone (top).

CHANNELWAY OBSERVATION

The channelway is now represented by the infill zone, and is clearly vein style. The channel is obviously very open, and with only minor evidence of strained borders is essentially extensional at this point.

CHEMICAL AWARENESS

Although more accurate mineral identification is required it seems clear that the fluid deposited cassiterite, magnetite and calcite whilst presumably effecting an iron-rich alteration of an essentially quartz-feldspar rock. Major alteration additions would include Fe and possibly Mg and Na. Major rock losses would include K and possibly Ca.

AUTHOR'S NOTE: **The albitic component of this rock was not recognised during hand specimen examination.**

PLATE 16 **MAGNETITE-CHLORITE-ALBITE ± EPIDOTE ALTERATION**
Leeuwpoort "C" tin mine, Rooiberg, Limpopo, South Africa.

PLATE 17

SILICA-PYRITE ± ANKERITE-CALCITE? ALTERATION.
Mt Charlotte gold mine, Kalgoorlie, Western Australia, Australia.

This specimen has been included to show the close-up detail of sulphide-silica alteration which is common in Archean greenstones. The gold is commonly associated with the alteration pyrite. The reader is also invited to inspect Plate 18 which although similar is perhaps more typical.

ALTERATION RECOGNITION

Alteration would be suspected from the changing mineralogy traversing from the host rock (bottom left) to the quartz vein (extreme top right hand corner).

ALTERATION POSITIONING

The required movement is at the eyeball level.

MINERAL IDENTIFICATION

The alteration zone consists of a prominent dull yellow mineral (pyrite) associated with grey-white glassy material (quartz?) and patches/spots of white/white-grey (calcite? ankerite?). The overall pale colour is very characteristic and striking as it usually occurs within dark (greenstone) rocks.

CHANGEOVER OBSERVATION (TRANSITIONAL ZONES)

The gradational alteration affect is seen along a diagonal traverse commencing from the bottom left in the least altered host rock. The host rock is unusually coarse-grained for the Kalgoorlie region and was originally a coarse-grained dolerite. The host rock is composed of pale white minerals which have a tendency to form lath-like crystals (plagioclase?) and dark ferromagnesian minerals which are rarely lath-like but mostly interstitial to the plagioclase (amphibole? biotite? chlorite?). Some vague pale green material is associated with some of the plagioclase (sericite?) together with minor reddish spots (hematite?) The host rock is still partially altered at the edge of the photograph but enough of the original texture/mineralogy is preserved to see the irregular but progressive changes towards the quartz vein.

The ferromagnesian minerals progressively contain increasing amounts of pyrite until their general position is completely occupied by rounded crystalline pyrite and pyrite clusters. The plagioclase crystals become increasingly less visible as they are replaced by quartz.

TEXTURAL OBSERVATION

The alteration process is progressively more texturally destructive as the intensity increases. The broad scale texture of the silica-pyrite zone bears a very broad-scale resemblance to the original dolerite, but has very little direct pseudomorphic character reflecting the original minerals.

CHANNELWAY INTERPRETATION

The increasing alteration intensity suggests a fluid channelway towards the top right and the quartz vein seems a good candidate. A small quartz vein (bottom left) also indicates a small-scale fracture channelway. This vein however shows little or no alteration halo and could belong to a different generation than that responsible for the main alteration.

CHEMICAL AWARENESS

The chemical changes from dolerite to silica-pyrite ore are fairly extreme, involving an introduction of silica, sulphur, gold and a loss of calcium, sodium (from feldspar) and probably magnesium (from the ferromagnesian minerals). The iron content might roughly balance out between ferromagnesian components – ankerite and pyrite. Some calcium may reappear as calcite.

PLATE 17 **SILICA-PYRITE ± ANKERITE-CALCITE? ALTERATION**
Mt Charlotte gold mine, Kalgoorlie, Western Australia, Australia.

◘ PLATE 18

CARBONATE-SILICA-PYRITE ALTERATION. Carbonation, Lake View gold mine, Kalgoorlie, Western Australia, Australia.

This specimen has been included to show the typical alteration style of many gold-related ores which occur within Archean greenstones. It is often referred to as carbonate alteration or carbonation and is associated with low, salinity carbon-dioxide-bearing fluids.

ALTERATION RECOGNITION

Alteration recognition is relatively simple with an obvious halo zone extending from a silica-sulphide-rich vein channel (top) into a dark host rock (bottom).

ALTERATION POSITIONING

The required movement is at eyeball level.

MINERAL IDENTIFICATION

The alteration zone is best seen from just over half to three quarters of the distance to the top of the plate above the white zone. It consists of striking white-pale pink ragged like bleb, lath and skeletal minerals (titanomagnetite -leucoxene - rutile) which are actually more obvious in the lower dark host rock zone. They are closely associated with finer-grained yellow blebs and spots (pyrite). These are set in a less distinct matrix consisting of pale buff material, darker buff-greenish areas, and paler glassy grey areas. It is very difficult to separate these as the colours vary. It would be enough however, to suspect the presence of at least two and possibly three other minerals. Quartz is a candidate considering the obvious siliceous nature of the main "vein" zone above. Sericite would also be suspected. The zone actually contains silica, sericite and a large amount of ankerite. The ankerite is probably the pale buff mineral which rarely approaches crystal form (left). The alteration is zoned and the silica content increases towards the top associated with a vague decrease (or different pattern) of sulphide distribution. At the hand lens scale most geologists would probably call this silica-sulphide alteration. Taken in geological context the "carbonation" label would immediately be suspected.

CHANGEOVER OBSERVATION (TRANSITIONAL ZONES)

The changeover zone is fairly sharp and to some extent obscured by the white buff zone across the middle of the plate. This is difficult to interpret. It could be the actual alteration front?, it could be a previous alteration of the dark host? or it could be a late overprint? This would need resolving by obtaining more specimens or possibly via microscopy. The acquisition of more specimens would reveal that it is certainly not part of an actual alteration front with the transition between the dark and siliceous layers being a consistent feature in much the same manner as seen in *Plate 17*. This example has been included to illustrate some of the traps awaiting observers when more than one stage is present. It is always prudent to look at the alteration zone in more than one place to establish consistency. This problem will be discussed more fully in a the next volume on overprinting. The fact remains that viewing the transition here is rather difficult (see *Plate 17*).

The continuity of the skeletal pale pink titanomagnetite from the dark zone into the altered zone suggests that altered rock is a derivative of the dark rock. The dark rock is one of the "Golden Mile" dolerites and was originally composed of pyroxene/amphibole (dark green) calcic plagioclase (white-glassy) and titanomagnetite. The dark ferromagnesian minerals have blurred outlines and are probably partially converted to chlorite (dark green). Although not readily discernible the alteration probably proceeded in much the same way as *Plate 17*, with plagioclase preferentially converted to silica, and pyrite nucleating in ferromagnesian mineral sites. A second changeover occurs towards the top of the plate where whiter silica predominates (see channelway discussion) .

TEXTURAL OBSERVATION

The alteration becomes progressively more texturally destructive with increasing silica although the main alteration zone displays a loose textural resemblance to the host rock. The upper zone, despite intensive. alteration, exhibits a distinct foliation and vague remnants of rock can be discerned which seem elongate and flattened. The texture is interpreted as ductile-high strain shear zone style deformation overprinted by silica alteration. Some narrow sulphide-rich zones traverse the top third of the plate and it is debatable whether or not these are an overprint shear/vein system or are directly related to the general silica-pyrite alteration. The prominent pyrite vein contains an unidentified dark mineral.

CHANNELWAY IDENTIFICATION

The fluids responsible for the bulk of the silica alteration fairly clearly gained access via the ductile shear system towards the top of the plate. It is difficult to judge just how much of this is silica alteration and how much is infill. From the texture, alteration silica is probably dominant which would be expected within this style of high strain zone access. Siliceous alteration is very texturally destructive and the blurring between infill and alteration components is a

PLATE 18 **CARBONATE-SILICA-PYRITE ALTERATION**
Carbonation Lake View gold mine, Kalgoorlie, Western Australia, Australia.

common problem of textural interpretation. A similar problem is encountered with common sulphides such as pyrite (***Plate 19***), pyrrhotite, chalcopyrite and sphalerite. This problem makes it difficult to say just how much of the small pyrite-rich vein (top) is infill. The dark crystals suggest infill whereas the spotty nature of much of the pyrite indicates alteration. It is probable that both are represented.

CHEMICAL AWARENESS

Depending on the position within the graded alteration zone, there are obviously additions of silica and/or sulphur. The disappearance of the ferromagnesian mineral suggests that iron and magnesium have either been redistributed or gone into the ore fluid. The pyrite could take up the iron and ankerite the magnesium. The ankerite suggests the presence of carbon dioxide. Accurate assessment even by analytical approaches would be difficult in the top section bearing in mind the difficulty of sorting infill from alteration and the uncertainty concerning whether or not the high stress zone formed before or accompanying the mineralising fluid. Most of the answers require the observer to move and look at more examples.

PLATE 19

PYRITE ALTERATION
(Sulphide alteration-replacement).
Mt Morgan gold mine (ore), Queensland, Australia.

This specimen has been included to demonstrate sulphide alteration within a brecciated host.

ALTERATION RECOGNITION

Alteration would be suspected due to the presence of a large dark fragment (middle-bottom to the right of frame) which is poorly defined and has potentially been taken over by sulphides (pyrite). This form of recognition presupposes that the observer has good "breccia eyes". Within the full hand specimen most fragments have been almost totally replaced, although there are some obvious fragments of silica veins with magnetite borders. Hints of sulphide veining are present within the plate (left hand side).

OBSERVATION POSITIONING

The specimen has been deliberately included to reinforce the single most fundamental rule in alteration study. There is ample evidence within the rock to suggest that a fragment has been broken, veined and severely altered. However, alteration is very well-advanced and the observer is advised to MOVE. In this instance the required movement would be substantive and of the order of tens of metres towards the margins of the Mt Morgan breccia pipe. Hopefully a less-altered zone could be located.

MINERAL IDENTIFICATION

The dominant yellow mineral (pyrite) is associated with both dark and paler glassy minerals. Both are extremely hard and tentatively identified as silica. The paler silica also appears as vague veinlets which are contained within the altered fragment. Although not visible the rock is also quite magnetic and magnetite is also present in fine-grained form. Given the pyrite dominance the alteration could be called pyritic. Sulphide alteration is extremely common and once again special terminology has been evolved to confuse geologists. For some unexplained reason when sulphide alteration predominates it is termed sulphide replacement. It is worth noting that it is just another form of alteration and should be investigated in the same manner as sericite, silica, or any other alteration assemblage.

CHANGEOVER OBSERVATIONS

The specimen is obviously a little unsuited to this form of observation owing to the advanced alteration. However, careful inspection of the border zones of the alleged channelways reveals numerous situations where the pyrite has "taken out" the host rock. The host rock mineralogy is very difficult to establish, and there seems a strong chance that it had already been affected by the underlying silica veining. It seems that silicification ± magnetite alteration occurred prior to sulphide introduction. A later, minute silica vein is also present (bottom right) and the rock clearly has an extensive paragenetic history. The grain size of the fragment suggests it may have been of a fine-grained sedimentary or igneous parentage. The observer should move to resolve this problem. The pyrite appears to have "attacked" all of the minerals present but obviously had difficulty with the paler silica. Initial alteration seems to be at the matrix level within the fine-grained rock.

TEXTURAL OBSERVATION (TRANSITIONAL ZONES)

The alteration is texturally destructive and at advanced stages leaves very little trace of the original rocks (see above).

CHANNELWAY IDENTIFICATION

Working on the principle that maximum concentration of the alteration assemblage probably denotes the permeability/access zone, it seems that the fluid entered the rock from both the top and left hand side. Potential fracture channelways are marked by wavy, vein-like trails of intense pyrite (left side). The channelways are best envisaged by viewing the plate from a distance or via slightly defocused eyes. Using this technique another very strongly altered fragment can just be envisaged occupying most of the top border zone.

This raises the question of channelway identification are there infill zones between fragments?, altered breccia matrix?, both? Pyrite always poses problems in this context as it habitually takes on a crystalline habit both as alteration and infill. The interpreted feeder zones contain very densely packed pyrite which is a little coarser-grained than some of the within-fragment pyrite. The writer would favour open space infill accompanied by intense sulphide alteration creating edge blurring effects. Again a less intensely altered specimen may resolve some of these questions.

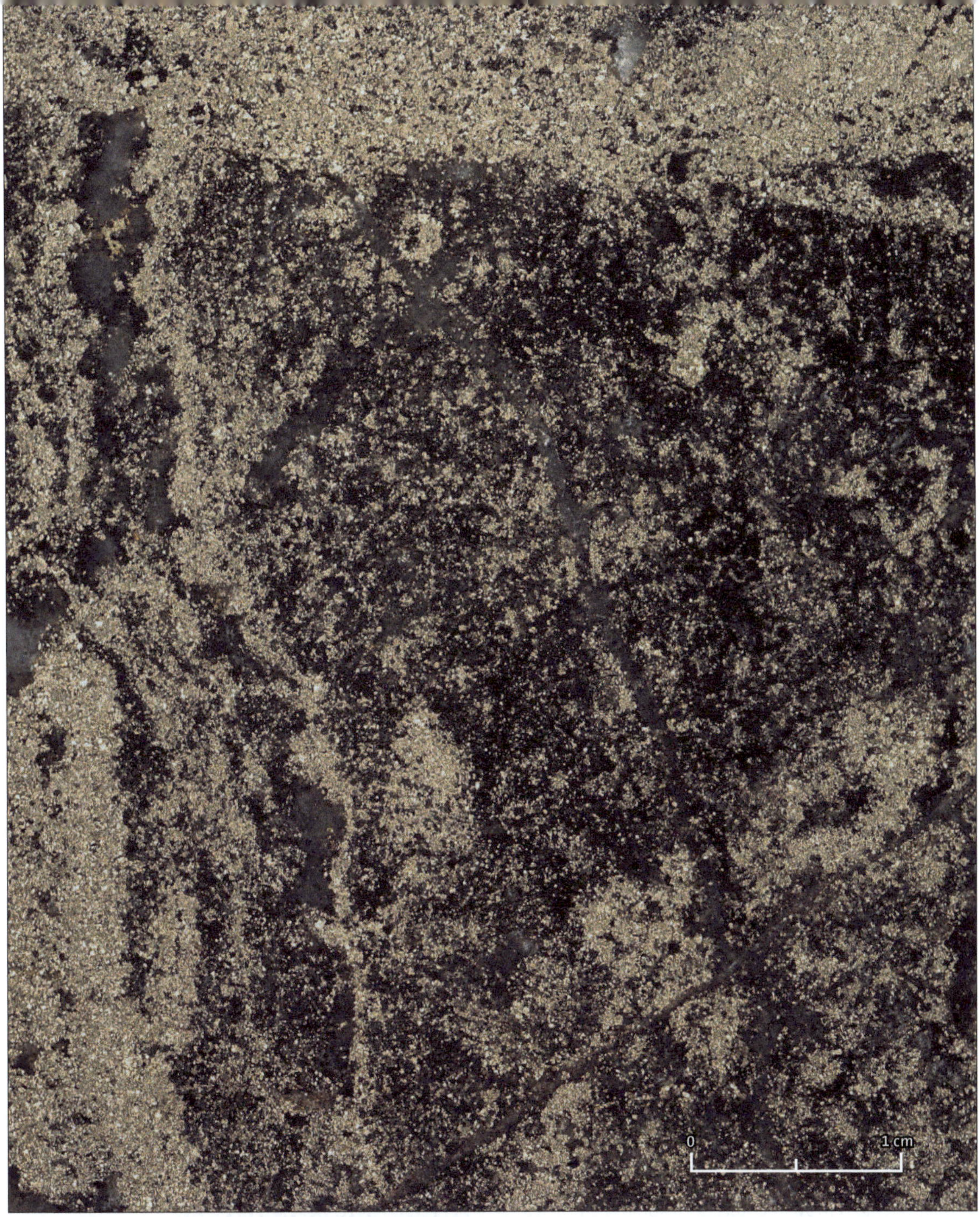

PLATE 19 **PYRITE ALTERATION (sulphide alteration-replacement)**
Mt Morgan gold mine (ore), Queensland, Australia.

CHEMICAL AWARENESS

Without accurate information on the host rock, it is a little difficult to assess chemical changes. It seems that the original rocks may have been converted to silica-magnetite prior to the introduction of pyrite. The first stage probably involved a loss of aluminium and gains in iron and silica, although an original host rock is required for confirmation. The sulphide stage suggests a major introduction of sulphur (and gold), possibly additional iron, and a loss of silica. Obviously more information is required on the fragment types (MOVE).

PLATE 20

PYRRHOTITE ALTERATION
(Sulphide alteration).
Watertank Hill gold mine, Mt Magnet, Western Australia, Australia.
(Specimen provided by G. N. Phillips).

This specimen has been included to illustrate both selective layer and sulphide alteration.

ALTERATION RECOGNITION

An alteration front would be suspected within this specimen from noting the facies change from pale brown (sulphides) to dark brown along the central layer. Sulphide presence always signals the possibility of sulphide alteration (replacement) and the suspicion would be enhanced by the presence of a potential feeder sulphide-rich vein system (left).

OBSERVATIONAL POSITIONING

Within the plate only eyeball movement is required. However the reader is again reminded that selective layer replacements/alterations are very common, often extensive, and easily confused with various species of sea floor/exhalative products. It is always wise to keep an open mind whilst seeking further evidence concerning change-over zones and/or feeder systems. Numerous serious mistakes have been made by hasty assignment of layer-controlled sulphides into either syngenetic or epigenetic categories.

MINERAL IDENTIFICATION

Accepting the alteration hypothesis, the most obvious product is a fine-grained pale-brown sulphide (pyrrhotite). The same mineral is present in other layers and to the left appears in a coarser-grained, irregular bleb style. It is also present within the vein-style fault related systems where it is accompanied by a darker dull yellow mineral (pyrite). The general alteration style would be called sulphide-replacement. the author would prefer to remove mental confusion by calling it a layer controlled pyrrhotite (or sulphide) alteration, but will have to admit the prospects of removing replacement from common parlance are remote!

CHANGEOVER OBSERVATIONS (TRANSITIONAL ZONES)

The composition of the host layer is difficult to decipher via hand lens observation owing to the small grain size. It is dominated by a dark mineral together with a paler grey mineral and occasional pale brown and pale pink spots. Some of the latter would be suspected as pyrrhotite related to the alteration. Within the hand specimen the dark mineral has a dull sheen and is intensely magnetic (magnetite). The paler grey mineral is probably silica although obviously this is a case for microscope support. Close inspection indicates that the magnetite component is under suspicion of being converted to sulphides. This is better seen on the specimen than in the plate. It is worth noting that sulphide replacement/alteration of magnetite is very common, and the general assumption is made that it provides a ready source of iron which promotes a reaction with sulphur-bearing fluids. The lower sulphide-rich layer also grades out into magnetite a few centimetres to the right of the frame. Observing a changeover into the coarser sulphide bleb material (centre left) is a little difficult owing to the effects of faulting. However, a general impression is given that the sulphides largely formed within a layer of different composition and texture.

TEXTURAL OBSERVATION

The alteration is texture retentive within the central layer with the sulphides taking the place of the pre-existing magnetite. The blebby sulphide style is less pseudomorphic in texture although still retains an elongation reflecting the host layers. It is also apparent from the plate that the rock has also been affected by another alteration. The massive white (quartz) area could well represent infill and there are many points where the layered metasediments seem to be converted to white-grey (silica?) texturally retentive alteration products (bottom and top left). The vein-style sulphide components give several indications that they cut both the infill and alteration silica (bottom left) and there is good reason to suspect an early brecciation/silica introduction prior to sulphide formation. It also seems probable that there are silicified rocks (centre-left) which are less prone to sulphide alteration than the normal silica-rich metasediments.

CHANNELWAY IDENTIFICATION

The vein sulphide channelway appears to essentially be a system of brittle-fault fracturing superimposed upon a similar, earlier system which has allowed entry of a silica-bearing fluid. Both the silica and sulphide systems have selectively replaced the layered metasediments although the latter is more extensive. Fluid ingress into the metasediments seems to be along the layer parallel fissility (S_O ?).

PLATE 20 **PYRRHOTITE ALTERATION, (Sulphide alteration), Watertank Hill gold mine, Mt Magnet, Western Australia, Australia. (Specimen provided by G. N. Phillips).**

CHEMICAL AWARENESS

The sulphide alteration consists principally of pyrrhotite which has "pseudomorphed" magnetite, presumably reflecting an addition of sulphur and a loss of oxygen from the original quartz-magnetite host rock.

PLATE 21

TOPAZ ALTERATION
Stewart Heads, Herberton tin district, Queensland, Australia.
(Specimen provided by G. W. Clarke).

This specimen has been selected specifically to illustrate the usual style and colour of topaz alteration.

ALTERATION RECOGNITION

Alteration within this rock is difficult to detect and at first glance it appears to be a pale coloured granite. However, valuable clues are given by the lack of sharp boundaries within the mineral assemblage. This is a common feature of considerable value for detecting alteration in granitoids. Both the white minerals and the darker components are rather blurred in character, especially with respect to their grain boundaries.

OBSERVATIONAL POSITIONING

If alteration is recognised the observer would naturally move to search for the host rock.

MINERAL IDENTIFICATION

The rock is composed of a white/cream mineral (topaz), a glassy grey mineral (quartz) and isolated darker patches ("muscovite"). Some of the larger white topaz zones contain small quartz granules. This is a common feature of topaz alteration of feldspar. The alteration could be called topazisation although most geologists prefer the less unwieldy term topaz alteration.

CHANGEOVER OBSERVATION (TRANSITIONAL ZONES)

Although the specimen has no fresh host rocks, the textural retention quickly leads the observer to suspect that the topaz is an alteration product of granite feldspar. The dark white-mica zones may also represent alteration of mica. Topaz is very common in tin systems where the accompanying white micas are lithium-rich.

TEXTURAL OBSERVATION

The host rocks are coarse-grained granites and there is a high degree of textural retention. The granite quartz is essentially unaltered.

CHANNELWAY IDENTIFICATION

The specimen gives little indication as to mode of fluid access. This actually forms part of a fault zone with quartz-topaz infill in a vein-style format. A close inspection suggests that many of the quartz grains are fractured and dislocated, suggesting a nearby zone of major fracture.

CHEMICAL AWARENESS

Topaz has substituted for granite feldspars and it follows that the rock will have lost K, Na and possibly Ca. Fluorine would form a major addition.

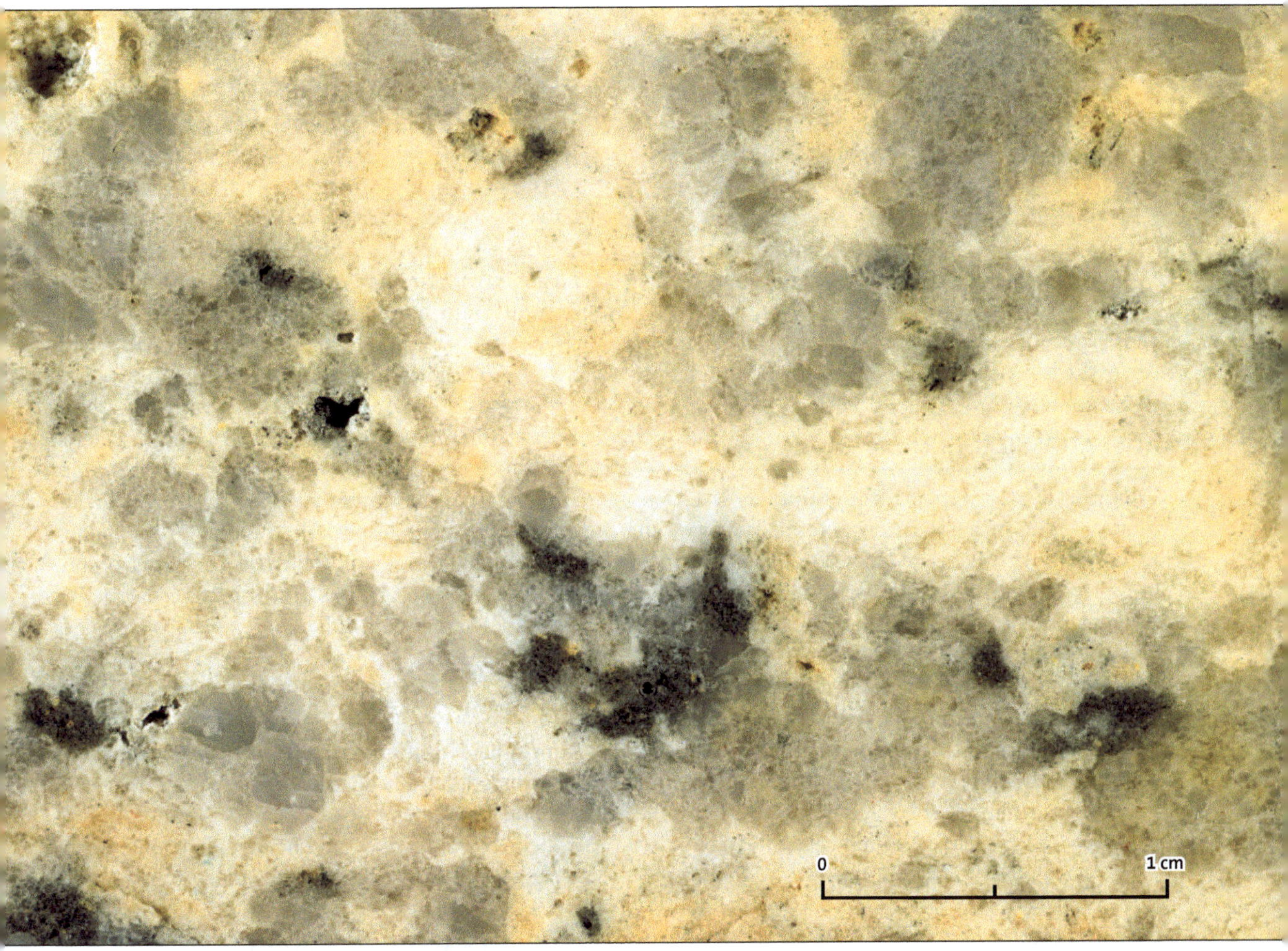

PLATE 21 **TOPAZ ALTERATION**
Stewart Heads, Herberton tin district, Queensland, Australia.
(Specimen provided by G. W. Clarke).

PLATE 22

GRANITE I "UNALTERED" GRANITE – WILD RIVER PLUTON
GRANITE II ALTERED GRANITE – NIGER CREEK PLUTON
Selective semi-pervasive alteration in granites, Herberton, Queensland, Australia. Specimens provided by G. W. Clarke.

These specimens have been selected to illustrate features which would suggest a semi-pervasive alteration in granitoids.

ALTERATION RECOGNITION

Background alteration within granitoids is a common guide to mineralisation potential and is well recognised within many tin-related systems. It is however difficult to recognise by eye and usually requires microscope work. The major visual clue is given by grain boundary blurring effects (mesostasis). Two granites have been presented here. Granite I is relatively unaltered and presents a general picture of fresh-looking grains with sharp grain boundaries. This contrasts directly with Granite II where the mineral grains appear rather variable in colour and have very poorly defined boundaries.

OBSERVATIONAL POSITIONING

Background semi-pervasive alteration is usually extensive and commonly effects entire plutons. In this sense no amount of physical movement will lead the observer into fresh rocks. A thin section is usually essential.

MINERAL IDENTIFICATION

Granite I is composed of potassium feldspar (pink), plagioclase approaching albite in composition (white), quartz (glassy grey) and biotite (black). Granite II has a similar composition but the mineral grains are more variable, potassium feldspar (varying shades of pink to white), plagioclase (white-pale green-yellowish), biotite (dark to less dark, note the ragged edges).

CHANGEOVER OBSERVATION (TRANSITIONAL ZONES)

It is not possible with this alteration style to move from fresh to altered rock. It is however possible to gain a few impressions by close inspection of grain boundaries and internal grain features. Granite II has suffered extensive development of secondary albite. This forms preferentially along potassium feldspar grain contacts, potassium feldspar-plagioclase contacts, and by enlarging existing perthite textures. The latter can just be seen within the potassium feldspars as pale white/pink anastomosing zones whilst the former form a myriad of small white grains in and around grain boundaries destroying any original sharp contacts. This is difficult to see but it may help the observer to assume that all small whitish-white green grains in the above plate are secondary albite. The specimen has also been affected by sericitic alteration which overprints the albitic alteration. This alteration is responsible for the variable colours within the plagioclase/albite (especially the grey-green-yellow effects). The biotite will have been chloritised and although not too obvious on this plate is normally expressed by a slightly duller, colouration and ragged grain boundaries.

TEXTURAL OBSERVATION

Almost by definition semi-pervasive background alteration is texturally retentive. However the degree of mineral-fluid interaction can vary substantially and mineral blurring (mesostasis) ranges from intense to very subtle. The above example would be regarded as moderately intense.

CHANNELWAY IDENTIFICATION

The channelways responsible for fluid access are not usually visible by eye. The alteration is thought to occur as fluids emerge at the final stages of granite crystallisation and become trapped in inter granular pore spaces and grain boundaries. Movement may also occur via this permeability system assisted by mineral cleavages, grain boundaries and minor microfracturing.

CHEMICAL AWARENESS

The chemical changes associated with this type of alteration are very subtle, and difficult to separate from major and trace element specialisation associated with normal granite fractionation. Detailed microscope and/or microprobe work is usually more effective for accurate delineation.

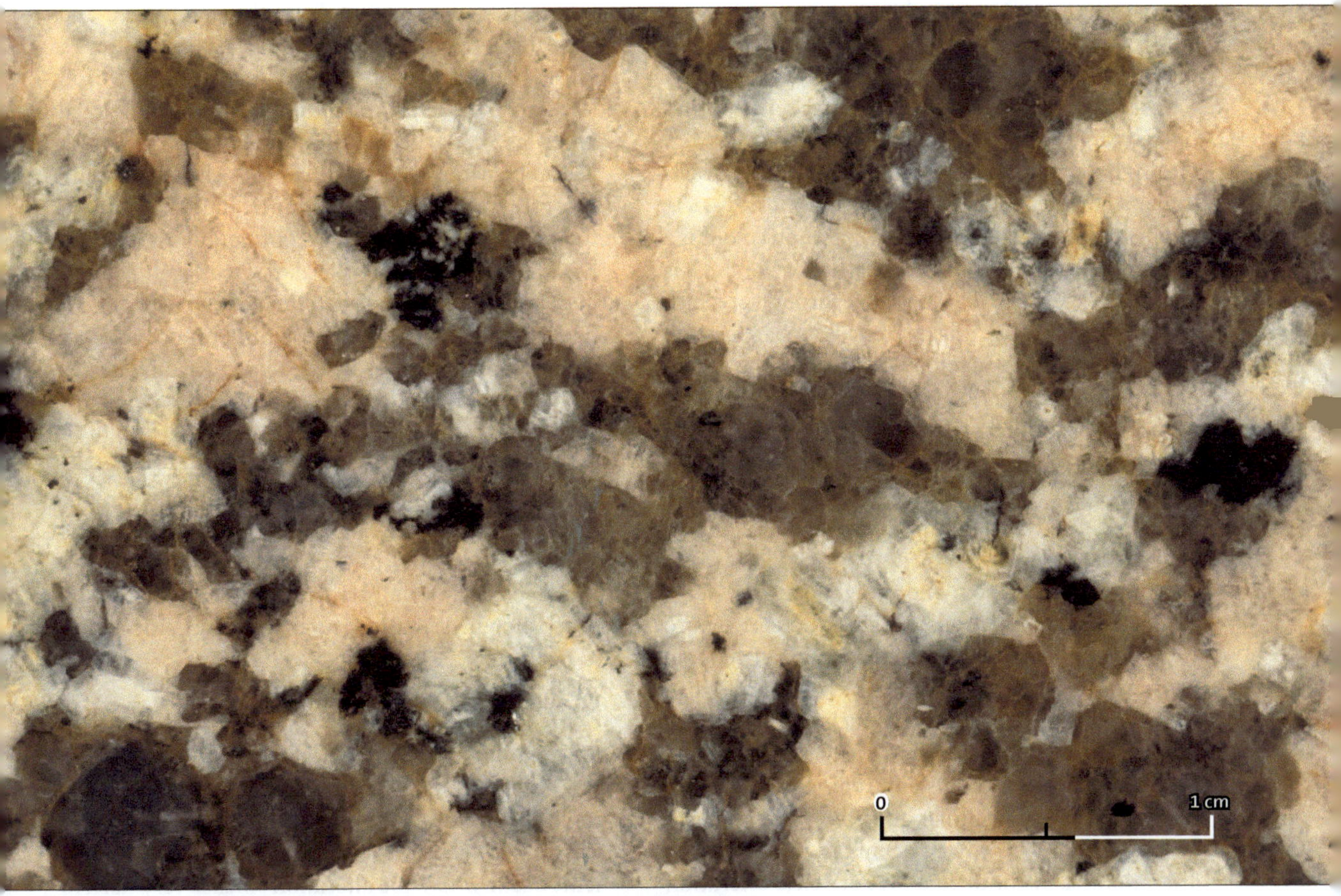

PLATE 22 **GRANITE I "UNALTERED" GRANITE – WILD RIVER PLUTON**
GRANITE II ALTERED GRANITE – NIGER CREEK PLUTON
Selective semi-pervasive alteration in granites Herberton, Queensland, Australia.
(Specimens provided by G. W. Clarke).

4 Alteration Minerals – Some General Observations

This section has been written as a "helpful hints" contribution. It is not intended to be comprehensive or a complete mineralogical expose. These are available in standard mineralogical works.

SERICITE – $K\,Al_2\,(Al\,Si_3\,O_{10})\,(OH)_2$

This term has no formal mineralogical definition and is best termed white, fine-grained potassium mica. It arose as a loose name given to fine-grained white mica which is easily visible under a microscope and has extended out as entrenched parlance in field identification. The reason for the loose usage is that even under the microscope the exact composition is not obtainable and total chemical definition requires detailed and time consuming follow up via X-ray diffraction and chemical analysis studies. In most cases the field or petrological observers are not prepared to go to these lengths. It is, however, mandatory for a full-scale research study. For casual treatment it is regarded as muscovite $K\,Al_2\,(Al\,Si_3\,O_{10})\,(OH)_2$.

From the hand specimen perspective sericite poses particular problems for inexperienced observers. Firstly it is usually too fine-grained to see clearly under a hand lens and secondly it assumes a variety of unexpected colours. The most common colour is a pale-buff greenish tinge not unlike muscovite, but it is commonly dark grey and can range through greenish tints to bright yellow. The darker colours are usually due to its strong association with fine-grained, dark quartz. The silica-sericite combination is very common with silica dominating near the fluid source and sericite taking over on the fringes. Sericitic alteration usually initiates within feldspars (especially calcic plagioclase) and the actual production of sericite also generates quartz.

Within tin, tungsten and molybdenum systems the varying combinations of silica and sericite are very common and the sericite is commonly coarse-grained (muscovite!). The mica frequently may contain additional components such as fluorine and lithium. To make alteration life even more confusing geologists working in this environment call the assemblage greisen. A similar quartz-sericite scenario within porphyry copper systems is called phyllic alteration! As mentioned above sericitic alteration preferentially attacks feldspars. However, there are many situations where white micas will actually alter all the minerals of the host rock to ultimately produce a mica product. This includes the quartz, and is particularly common in tin systems. Sericite infill is a common corollary of sericite alteration and can be distinguished as the infill component via a coarser grain size.

CHLORITE

Chlorite belongs to a large group of platy, monoclinic minerals. The general formula is $(Mg, Fe^{2+}, Fe^{3+})_6\,Al\,Si_3\,O_{10}\,(OH)_8$. Chlorite alteration is very common and like sericite can range from a selective attack centred upon ferromagnesian components or take out the entire rock. The term chlorite embraces a wide range of compositional variations although iron-rich varieties are common. These are impossible to differentiate by eye, difficult to separate beneath a microscope, and identification of the mineralogical composition requires both probe and X-ray Diffraction work. Mixed layer arrangements are common within individual "chlorites". Chlorite identification poses problems for beginners who are unfamiliar with chlorite colours. Most alteration chlorite is much darker in colour than might be anticipated. It commonly appears to be black and the touch of very dark green is difficult to discern. The problem is accentuated in partial alteration situations where chlorite commonly occurs as an alteration of black ferromagnesian minerals such as biotite or hornblende.

EPIDOTE $Ca_2\,(Al\,Fe)_3\,Si_3\,O_{12}\,(OH)$

Epidote and members of the epidote group are common alteration minerals. Recognition of epidote is usually easy owing to the characteristic pistachio green colour. It usually replaces plagioclase feldspars or ferromagnesian minerals in a patchy manner, and is a common associate of chlorite.

QUARTZ SiO_2

Siliceous alteration (silicification) is extremely widespread and usually presents very few problems to most observers. A simple hardness test usually serves to confirm suspicions of silica content. Minor problems occur where the alteration silica is very dark in colour and there are situations where pink and buff silica colours resemble carbonates. The alteration covers a very wide range of temperature conditions. Silica alteration is commonly very texture destructive and there are major difficulties in recognising the original rocks, and in separating silica infill from silica alteration. This may lead to either incorrect or more commonly, no recognition of fluid channelways. For this reason many breccias remain undetected and the breccia type consequently unrecognised. Many "cherts" ultimately prove to be silicified breccias. In some cases the silica infill may be slightly different in colour or texture to the alteration silica, and differences may be more obvious at the fringes of the altered zone. Although not a universal guideline many occurrences of very dark silica actually contain very fine-grained sulphides. This is particularly true in epithermal systems, but any dark silica should be inspected carefully for sulphide content.

FELDSPARS

POTASSIUM FELDSPAR – $K\,Al\,Si_3\,O_8$

Potassium feldspar is a relatively common alteration product in the sense that most geologists will encounter "K-spar" alteration as part of the potassic alteration zone within porphyry copper systems. It usually results from the alteration of pre-existing plagiocase and potassium feldspar (commonly perthitic) and is often quite subtle and difficult to see. The alteration feldspars range in colour from pale pink (flesh pink) to white and are commonly only partial alteration products occurring around the rims of the original feldspars. Even the more colourful pinker varieties are difficult to pick from the general feldspars of the host rock and thin section follow up usually reveals a lot more alteration than originally suspected. The association with secondary biotite (especially narrow wispy dark microveins) is especially helpful. Many zones of potassic alteration in porphyry copper systems are actually dominated by secondary biotite. This is especially true where the host rocks are towards the ferromagnesian end of the spectrum within many of the island arc style systems. Potassium feldspar alteration is not restricted to porphyry copper systems and is rarely well-represented in early phases of tin systems. In general terms it is regarded as a high temperature alteration product (400–600 °C) associated with saline magmatic fluids.

The author has seen one situation at the base of a tungsten quartz vein system where potassium feldspar had completely replaced all the minerals of the host granite in an irregular vein-like format. However, no obvious channelway could be seen. Close inspection revealed that the entire zone contained minute infill spots. The infill was made up of small crystals of potassium feldspar identical to that of the alteration. It seemed as if initial dissolution (of quartz?) had occurred providing a sponge like alteration channel. A similar phenomenon is seen in many tin-related albitic rocks.

ALBITE $Na\,Al\,Si_3\,O_8$

Albite is a very common alteration mineral especially in tin-related systems. It also occurs associated with some porphyry copper environments and within Australia is particularly prolific in the Cloncurry region. Within the latter it is regionally distributed and held by many to relate to original evaporitic sequences. Albite is commonly difficult to recognise as it is mostly fine-grained and varies substantially in colour (white, cream, flesh pink, pink). The colour variation results in confusion with K-feldspar, hematite, silica and carbonate. In many cases the suspected presence of alteration albite can only be confirmed by continuing petrological work. This is particularly so within fine-grained sediments. Albitic alteration is also common within rare-metal pegmatite deposits.

ADULARIA $K\,Al\,Si_3\,O_8$

Adularia is a moderate-to low-temperature triclinic mineral of the alkali feldspar group. It is a particular form of K-feldspar and is relatively uncommon as an alteration product. It assumes importance within epithermal systems as part of the vertical zonation system. It is usually white (rarely pink) and can be very difficult to recognise especially when fine-grained. Within epithermal systems it is very common as an infill where it is recognisable from the crystal form (pseudo-orthorhombic).

TOURMALINE

Tourmaline actually belongs to a group of minerals of which schorl is the most common in hydrothermal systems – $Na\,Fe_3{}^{3+}\,Al_6\,(Si_6\,O_{18})(BO_3)_3\,(OH)_4$. Tourmaline occurs prolifically within tin and pegmatite systems and is by no means uncommon in a more general sense. Significant amounts are present in many of the South American porphyry copper systems. Alteration tourmaline assumes a variety of forms and boron-rich fluids seem to have an ability to be amazingly penetrative, without much sign of fluid channelway access. Isolated clots and rosettes of tourmaline may occur metres away from the main feeder channels, and the author has seen several examples of fine layering (including cross bedding) selectively replaced over substantive distances. Recognition is relatively easy from the striated prismatic crystal form although fine-grained examples require petrological confirmation. Tourmaline alteration is commonly associated with silicification and argillisation.

CLAY-ARGILLIC ALTERATION

Clay alteration can only be described as a nightmare for both novice and experienced geologists. Most geologists have no problem in recognising clay, but all have severe problems in identifying the type of clay and whether or not it is derived from the weathering process or via some deeper-seated hydrothermal system. Surface clay alteration is almost inevitable in sulphide-rich systems. The oxidising sulphides quickly generate a low temperature "acid bath", and clay production predominates within the alumino-silicate host rocks. This style is readily suspected by the presence of substantial iron staining (limonite) and obvious leached outcrops, denoting the former presence of sulphides (pyrite, pyrrhotite). The clay alteration usually extends down to the water table. The problem still remains of whether or not there was any original clay alteration prior to weathering. In most cases this proves to be not the case but the problem is usually only resolved by deeper mining activities.

Identification of clay and clay-related alteration minerals is effectively impossible by eye, very difficult by microscope, and requires specialised equipment such as a short wave infrared spectrometer. Coarser-grained pyrophyllite $Al_2Si_4O_{10}(OH)_2$ can be recognised as it resembles a soft greasy muscovite, but in general the beginner is advised to call for experienced back up!

HEMATITE Fe_2O_3

Hematite is a very common alteration product and is quickly suspected from the pronounced red colouration of the alteration zone. In most cases it occurs as a very fine dusting throughout the altered minerals (especially feldspar) and requires petrographic support for confirmation. Most red colourations are due to fine-grained hematite, although there are situations where pink/red carbonate, albite or potassium feldspar cause red colours to predominate. In this sense a little caution is advised with pink/red alteration products.

SULPHIDES

Sulphides are extremely common alteration products (often referred to as sulphide replacement). Pyrite, pyrrhotite, chalcopyrite, sphalerite and arsenopyrite are prolific alteration products and in some cases can totally take over the host rock. Rocks and minerals containing iron or calcium are highly favoured, although by no means exclusively. Pyrite and arsenopyrite readily assume euhedral crystal shapes. This tendency together with the tendency for pyrrhotite and chalcopyrite to avoid euhedral shapes in either infill or alteration mode, poses many problems for the observer. These problems usually make the separation of infill from alteration very difficult. The common chalcopyrite/pyrrhotite combination is especially problematic as infill and alteration are commonly texturally identical. The presence of associated quartz crystals invariably denotes an infill zone, but even with this guide it may prove impossible to tell where infill passes into alteration.

TOPAZ $Al_2(SiO_4)(F, OH)_2$

Topaz is extremely rare as an alteration (or infill) mineral and is virtually restricted to selected tin systems. Recognition is difficult and the most common form is as a creamy-white mineral occupying the sites of feldspar within "granite" rocks. In this format it is easily overlooked and confused with quartz. In most instances the quartz is darker and glassier. The addition of topaz commonly results in some very hard compact rocks which may prove troublesome in any consequent crushing or grinding operation. In general terms topaz signifies a high temperature system and signals the presence of fluorine within the fluid.

CARBONATES – CALCITE $CaCO_3$, ANKERITE $Ca(Mg, Fe, Mn)(CO_3)_2$, SIDERITE $FeCO_3$, DOLOMITE $CaMg(CO_3)_2$

Carbonates of various species are very common alteration products and range from very easy to detect to rather obscure. They tend to reach maximum development in rocks which contain significant amounts of original carbonate, but also occur in rocks which have significant amounts of calcium-rich plagioclase especially in association with propylitic alteration styles. They are always a potential candidate in fluid systems containing CO_2 and for this reason appear in abundance in association with greenstone gold terrains. In these regions the alteration style is commonly called carbonation with sizeable carbonate (ankerite) crystals appearing in the pale alteration halos. Carbonate alteration takes on many forms and there are a whole class of ore deposits simply labelled carbonate-replacement deposits where carbonates (usually iron- and magnesium-rich) occur in abundance. It is worth noting that many carbonate species easily form perfect crystal shapes in alteration zones and can easily be confused with infill. They also form perfect sites for any later phase sulphide alteration. Sulphide alteration is thus commonly be found overprinting previously formed carbonates. With the diversity of Fe, Mg, Ca combinations they assume a wide range of colours from chocolate brown iron-rich members (siderite) through buff (siderite, ferroan-dolomite) to white (calcite) colours. The addition of minor trace elements further adds to the colour range. Pink carbonates give particular early recognition problems as they are easily confused with a wide range of pink alteration products (K-feldspar, albite, hematite, silica). Because most late-stage carbonate veins are practically devoid of alteration halos, it is easy to form the opinion that carbonates will not easily form as alteration products in standard aluminosilicate rocks (shales, sandstones). The author has however, seen several examples of carbonates totally replacing quartz!

MAGNETITE Fe_3O_4

Magnetite is a common alteration mineral and is commonly found as a moderate- to high-temperature alteration mineral associated with ferromagnesian host rocks. It is also common within skarn systems (which are not treated in this text). Identification is commonly remarkably difficult in finer-grained examples where the obvious hand magnet test is not always performed by the observer. The examples in ***Plates 16 and 19*** both escaped detection during initial inspection.

GENERAL OBSERVATION

Within granitic rocks it is worth noting that many alteration change inwards toward the channelways in almost the same manner. This relates to the general truth that calcium- and iron-rich minerals are particularly sensitive to reaction with fluids, whilst quartz is generally resistant. Most of these alterations are texturally retentive in their outer zones with the original textures becoming progressively destroyed approaching the main fluid channel. Plagioclase is commonly the first mineral to show signs of change (mostly to clay products). This is closely followed by the ferromagnesian components (mostly to chlorite) although this is more difficult to see by eye and is often just noted as general blurring of originally sharp grain boundaries. This is again closely followed by the potassium feldspars which tend to resist total annihilation further into the alteration zone. Quartz is the last to show any visual change but eventually even the sharp edge of the quartz grains begin to blur. The above sequencing holds a general truth for chlorite, sericite, topaz, tourmaline and many sulphide styles. Advanced alteration will eventually "take out" the resistant quartz and in many cases result in a monomineralic product (chlorite, sericite, sulphide).

This preferential sequencing yields many clues to the nature of the fluid and it is essential that observers cultivate the habit of working inwards on a mineral by mineral basis. In most cases the changes are readily observed under a hand lens. The only real barrier to this approach is in very fine-grained sediments where the minerals are too small for hand lens resolution.

5 Assessment Sheet

This text commenced by saying that it took the author many years to develop good habits when confronted with alteration, and although most practical approaches and principles outlined here may seem elementary, many of them eluded the author for some time. The check list below has been included as a final reminder and there are no objections to anyone who would like to copy it for field use. An apology is perhaps due for the low key tone of the text but the author can still recall the many problems of the early years and sincerely hopes this will be of use. A full understanding of alteration is a valuable professional skill, and hopefully future texts on overprinting and paragenesis will contribute further.

THE LOOK AND THINK PROCESS

This is especially designed for those with the "morning after" syndrome which is very common in fieldwork situations.

1. Pick up hand lens	
2. What has been altered? **(The move principle)**	
3. What are the new minerals? **(Mineral typing or at least noting they exist as red ones etc.)**	
4. What minerals have changed to new minerals? **(What is happening in detail?)**	
5. Is it texturally retentive or destructive? **(Total wipe-out or just a nibble)**	
6a. How did the fluids gain access? **(A most important question which involves 6b – vein, breccia etc.)**	
6b. Do we need to sort alteration from infill? **(In most cases the answer is yes – this is the most common source of misconception – which is which?)**	
7. Is there more than one alteration involved? **(Paragenesis – involves close observation with brain in gear)**	
8. What are the main chemical changes? **(As dictated by the new minerals – clues on fluid chemistry)**	

6 References

Guilbert, J. M., and Park, C. F. Jr., 1986. Geology of Ore Deposits. Freeman, New York, 985 p.

FROG

This illustration might at first sight be considered as irrelevant. It is however a North Queensland green tree frog and serves to show just how difficult it can be to detect obscure alteration processes. A reliable source indicates that it was once a handsome prince.

Volume 3 – Overprinting Textures

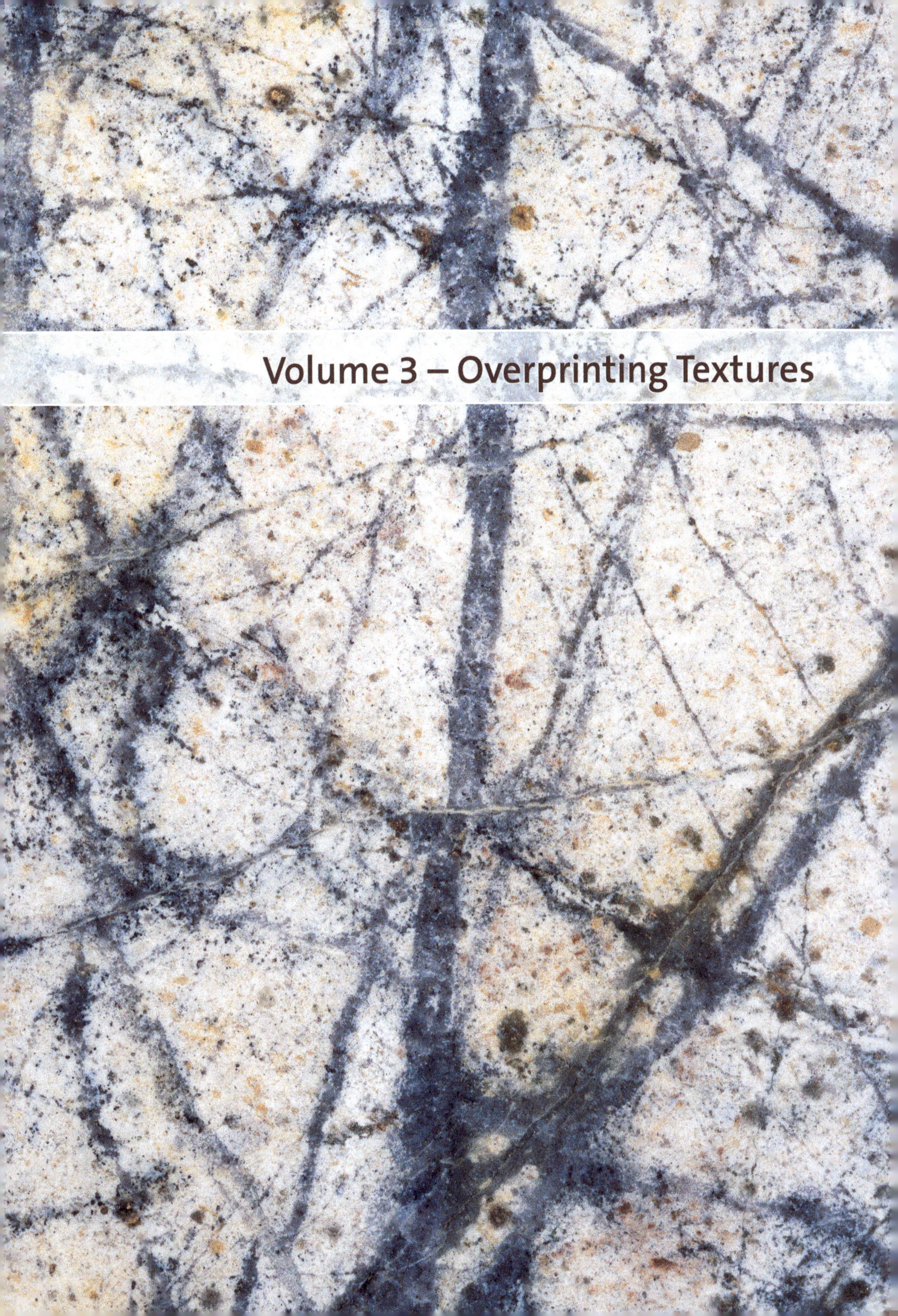

Volume Front Page

Specimen from Escondida copper mine, Chile. The specimen has been chosen to catch the eye of wealthy browsing customers. It is from a porphyry copper system, with at least three overprinting vein stockworks readily visible. Unfortunately it has been subjected to argillic alteration, and is from the zone of supergene enrichment. Hence most of the dark minerals present are chalcocite/covellite after original sulphides (chalcopyrite/pyrite?). The oxidation/secondary enrichment obscures the primary infill/alteration features, and a deeper primary ore specimen would be better for paragenetic/overprinting work. It does however illustrate the complexity in porphyry systems, and makes a nice photograph.

1 Introduction

Many of our concepts concerning the nature of hydrothermaal systems and their relationship with rock fracturing arise from the common observation that hydrothermal fluids seem to precipitate different mineral assemblages at different times. Practically every mineral deposit will provide an example of crosscutting vein, which are conventionally interpreted as two (or a single) generations of hydrothermal fluids with an intervening period of rock fracturing. The crosscutting relationship is an example of an overprinting texture and the implied timing relationship allows the two veins to be catalogued as a temporal sequence, such as Stage 1 and Stage 2. The process of constructing a time sequence is referred to as establishing the paragenetic sequence.

Nearly every deposit exhibits some three to five stages of development, which reflect the close association between long-lived hydrothermal systems and associated tectonism. Some deposits of the epithermal or porphyry style may exhibit numerous stages, reflecting very active rock fracturing (usually due to the boiling of fluids) which continues intermittently throughout the history of hydrothermal fluid flow.

An understanding of overprinting textures and their paragenetic implications is the basis for all ore deposit studies. It would be fair to state that until this aspect is well-understood, any further studies regarding such items as fluid inclusions, stable or radiogenic isotopes, alteration geochemistry and fluid modelling, run the very grave risk of setting off from an incorrect premise. This situation is much more common than most of academia is prepared to admit! From a practical perspective the comprehension of paragenetic sequencing is equally important. Firstly, it provides the early information which serves to give the exploration team some concept of the type of hydrothermal system under investigation, and it will come as no surprise to exploration managers to learn just how many times this has been misinterpreted. Secondly, within any multiphase system it is very important to understand which phases are actually ore and to comprehend their spatial distribution.

From every perspective it becomes a prime geological requirement to fully delineate and interpret the many facets involved concerning overprinting and paragenetic sequencing. The textures provide data concerning structural environment, fluid composition, structural controls, ore distribution, mineral zoning, temporal and thermal history of fluid flow, and a host of related subtleties that sharpen comprehension and hence exploration targeting and continuing research.
Given the above it would be natural to assume that both the academic teaching system and the mineral exploration community would stress the importance of this aspect. Regretfully the world is a busy place, with academia having a large amount to fit into a three or four year programme, and industry having to find ore quickly. Most geologists would conclude that there is room for improvement, and this small text is offered in that spirit. The rocks will speak, but it always takes time to learn a foreign language.

Fortunately within the domain of overprinting and paragenesis, practically all the major observations are to be made at the outcrop and hand specimen scale. It is much easier to sort things out at the rock face scale, than beneath a microscope. Rock slabs and drill core are also prime treasures, as both provide clean smooth surfaces enabling close observation. All of the above are readily available in most exploration and mine environments, and when used in combination are powerful weapons for resolving overprinting and hence sequencing problems. The rock slab is definitely the most underutilised, and the tendency to leap from rock face (or drill core) to the microscope frequently turns an easily soluble problem into confusing situations.

In keeping with the rest of this series, the text is aimed at observations which can be made either by eye or with a hand lens. The plates have been selected to show a range of mineralisation styles, and plate descriptions have been written at length to indicate the approach and thought processes involved.

2 Overprinting Criteria

2.1 General

Within the context of this book the term overprinting refers to any textural observation which can be utilised to infer that one mineral (or group of minerals) has been deposited later than another. The aim in all overprinting observations, is to establish a sequence of mineral deposition which is referred to as a paragenetic study. Overprinting observations are thus the building blocks which produce a paragenesis. The entire process is absolutely fundamental to anyone who is trying to establish exactly "what is going on" within any form of hydrothermal mineralising system.

Overprinting criteria are very simple in theory, but tend to be rather complex in practice. There are numerous small traps awaiting the unwary or inexperienced observer, and as previously stated a good grounding in the difference between infill and alteration concepts is essential. In practice a range of techniques are utilised, which range from first to fourth order depending upon the degree of interpretation.

2.2 First Order Criteria (Confidence Building)

(a) Mineral Superimposition

Hydrothermal fluids traversing "open" channelways may precipitate minerals (infill) and where one mineral can be seen to have nucleated upon another, an obvious timing (overprinting) relationship can be inferred. Numerous examples are given in the first section of this book, and others are displayed in this text. (*Plates* 1–3).

(b) Structural Superimposition

Rock or mineral failure occurring after a period of hydrothermal deposition, will fracture and/or fragment previously formed minerals. This common situation provides a range of new (overprinting) channelways for fluid flow and deposition. The net result provides the two most common criteria utilised to separate one period of mineral deposition from another:

(i) Crosscutting vein systems (each with separate infill alteration components). (*Plates* 4–*11*).
(ii) Fragments (breccia) of early mineralisation contained within the new mineralisation (*Plates 12–15*).

2.3 Second Order Criteria (Suspicion Arousing)

Any structure which has acted as a conduit for a hydrothermal fluid and subsequent deposition, is very prone to reactivation during a subsequent rock fracturing event. In simple terms faults, veins, breccia zones tend to form the weak links in the chain which keeps on breaking in the same place. It is very common for a succession of mineralising events to occupy the same zone of structural weakness, and the simple structural superimposition criteria presented above may become rather obscure. Given that each influx of fluid will result in a new infill assemblage together with alteration of either preexisting alteration, or previously unaltered host rocks, it is not surprising that complex situations may occur. Observations that raise suspicions of multistage overprinting include the structural superimposition criteria mentioned above, as well as less obvious situations such as:

(a) Mismatches Between Alteration and Infill Components (Things Out of Proportion). Including:

(i) Small channelway structures (usually veins) traversing excessively larger areas of alteration.
(ii) Widespread alteration which does not appear to increase significantly (or at all) as it approaches a sizable channelway structure (vein, shear, "triangular vug").
(iii) Veins (usually first noted as infill structures) which although centrally located around alteration within the majority of the rock, "wander across" the alteration zone to produce grossly asymmetric situations.
(iv) Asymmetric mismatching also occurs in infill sequences. Although there is no inherent reason why minerals should deposit symmetrically on

each side of an opening, there are many situations where the observer intuitively suspects this should occur within the style of specimen being examined. An extra layer on one side is always suspicious.

(b) Inconsistent Alteration Configurations

There are many variations to this criterion, but it is very common for an alteration zone to overprint pre-existing alteration. If the second alteration completely overprints the first at some points, but only partially overprints at others, the observer will be presented with different zonation patterns at different points. Thus working from the host rock inwards towards the suspected channel zone will give different results at different points.

(c) Alteration of Alteration

To some extent this is a variation on the above, but it is quite common to see alteration products which retain the texture of previously altered rock.

Examples of second order (suspicion arousing) criteria are given throughout the plates, with some selected situations being illustrated in ***Plates 16–22***.

2.4 Third Order Criteria (Indirect Overprinting)

Almost any box of core, or large rock slab, situated in the marginal zones of structurally controlled ore will exhibit isolated veinlets with alteration/infill assemblages which are obviously different to each other. The same situation will occur with isolated veins/veinlets located in two different places. The presence of two different assemblages suggests two different stages, but obviously gives no timing information.

Another very common suspicion arousing indicator is a sudden change of crystallisation style within an apparently continuous sequential infill sequence. A sudden change from coarse crystalline comb quartz to finely layered "colloform" styles would be a common example (***Plate 1***)

2.5 Fourth Order Criteria (Indirect Overprinting Temperature Inference)

Where the above isolated veinlet situation occurs, it is of considerable assistance if the observer has some concept of high temperature versus low temperature assemblages. Most paragenetic sequences represent declining fluid temperature sequences. A biotite-rich assemblage would normally indicate much higher temperature than one which was predominantly clay. It does not necessarily follow that the biotite is earlier, but it is a good working hypothesis. This style of criteria needs to be used with caution, and hence the fourth order grouping.

3 Overprinting Textures

3.1 Broad Scale Perspectives

Overprinting textures inherently refer to interpretations regarding the relative time sequencing of hydrothermal events. However, it is important to perceive that the time sequencing observations also offer the opportunity to look at each stage separately (or in combination) to establish a wider range of geological parameters concerning the structural and fluid history of an individual ore deposit. Admittedly the amount of visual information that can be extracted relates to the knowledge and experience of the observer, but the opportunity remains to consider evidence relating to ore genesis from a range of interrelated perspectives. These include:

(a) Structural evolution as reflected by the style(s) of fracturing. This could be restated as understanding the development of fluid channelways, and hence structural controls.
(b) Chemical composition and evolution of the ore fluids as reflected by the composition of the infill minerals, and their associated alteration assemblages.
(c) Factors relating to physical parameters of the ore fluids. Temperature indicators are readily available from the depositional sequence. Parameters relating to pressure and pressure variations are present in textures that can be related to processes indicative of hydraulic fracture, hydraulic brecciation and boiling. In many cases even the crystal sizes and shapes can be tentatively related to possible cooling rates, fluid saturation, or phase separations. This is especially true with quartz textures (*Dong et al., 1995*).

3.2 Recording Observations

It is always wise to record observations, and this is particularly true in overprinting studies. Many situations are ambiguous and/or subtle, and it is crucial to look carefully and not to try to force conclusions from unclear examples. It is quite surprising just how many apparently clear/simple situations are in fact complex. A useful practical suggestion, is to construct a table as indicated below.

Stage (Name)	Infill*	Alteration (Rock type altered)	Structural Style	Comment
"Chlorite"	Quartz, chlorite, pyrite	Chlorite, minor sericite? and carbonate (Basalt) Irregular edges	Small vein 2–3 cm wide blurred.	Distinction between infill and alteration blurred.
Pink vein	Pale pink minerals. Sharp edge against alteration	Darker pink (Granite) expressed by pinking of feldspars	Large vein 6–8 cm wide. Infill predominates. No brecciation. High fluid pressure – jacking fracture open?	Suspect pink infill and alteration are albite or K-spar? Cuts small chlorite vein, with no apparent effect?

* Infill. It is very rare that the sequential infill order can be easily observed on one sample. If relationships are apparent they can be recorded. For example – quartz overprinted by chlorite, or quartz (chlorite, pyrite, uncertain timing) etc.

4 First Order Criteria – Mineral Superimposition

Sequential Infill

Textures and problems. Plates 1–3

Infill textures provide extremely valuable information concerning the sequencing of mineral deposition. Within a single stage of deposition it is very common for minerals to precipitate as crystals or layers. Early formed crystals or layers become nucleation zones for succeeding precipitation and in ideal situations it is relatively easy to determine a sequential sequence. This is most easily observed in large-scale cavities which are only partially infilled, and many geologists would be familiar with spectacular examples which occur in Mississippi Valley style environments. When the fluid space is completely infilled, sequential deductions are a little more difficult, but can generally be achieved with only minor problems. Criteria for recognising infill are presented in the first part of this book, and it is worth noting that this is a much neglected skill.

Within veins and breccia cavities crystals often nucleate as isolated individuals, rather than as the continuous layers which are so often depicted in textbooks. In this situation it can become very difficult to find examples to clearly illustrate the complete relationships between the three to five minerals which have precipitated. The problem can usually be resolved by inspecting a large number of samples of the particular stage and slowly teasing out the critical "overgrowth" relationships. Observers should not expect to discern the full sequence from one specimen, and in many cases will be left with a few uncertainties concerning depositional timing between two or more of the minerals. Common minerals which give particular problems are the chalcopyrite/pyrrhotite pair. These commonly precipitate together and neither likes to form crystals. It is also quite possible for minerals to precipitate simultaneously or continuously throughout. This can obviously produce contradictory overprinting results.

Theoretically the simultaneous or continuous precipitation problem could make the concept of sequential overprinting of limited value or impossible to apply as a practical reality. However from twenty years of experience, the writer can assure the reader that in the vast majority of cases minerals normally precipitate in a simple sequential manner, which is readily deduced with a working knowledge of the principles of infill recognition.

A much more serious problem arises when it is realised that a void space, such as a large breccia vug or an open fissure, may receive mineral precipitation over a long period of time and there is no guarantee that the sequence of precipitation deduced actually represents a single stage. Obviously the larger the void space the more the opportunity exists for this to occur. Although refracturing may have occurred and a new stage been introduced, it is commonly not reflected within the void space (it is difficult to fracture a hole!).

There are several ways of resolving a problem of this nature but the first step is to be aware of the possibility. An early warning signal is usually given by the textural and or mineralogical changes within the sequence. The inner regions of the vug may record a sharp textural change in crystal size. Alternatively the entire mineralogy may suddenly change. In many cases both textural and mineralogical changes occur which raise suspicion. With a little knowledge of temperatures of deposition, it may be obvious that very high-temperature products are suddenly succeeded by low-temperature assemblages.

A more conclusive resolution is almost always achieved by inspecting more specimens. If the suspect portion of the infill is in fact a different stage, it will appear as a distinct entity elsewhere within the new fracturing, and it is usually possible to find it cross-cutting the previously-formed materials.

PLATE I

SEQUENTIAL INFILL
St. Patrick tin mine, near Irvinebank, Queensland, Australia.

This specimen has been selected to illustrate the problems of sequential infill. It will also serve as a good example of infill textures.

The specimen is composed almost entirely of quartz which shows considerable variation in both colour and style of crystallisation. The bulk of the quartz is coarsely crystalline (comb quartz) although there are several episodes of fine-grained deposition producing a layered effect. This is prominent in the mid plate region. Most of the quartz is in layered format, but there is a distinct textural change in the centre of the vein, where coarse-grained white crystals have nucleated in a relatively random manner. The final stage of deposition is cassiterite (black) which exhibits some magnificent triangular textures as it fills in, between quartz crystals.

Most of the above is readily visible but the real question is: "How many stages of deposition are represented here?" It seems that a large cavity has been progressively filled. The definition of a stage leaves a little to be desired, but is generally taken as a group of mineral precipitates which have been deposited during a continuous fluid flow. This is impossible to establish by scientific observation, and in practice a stage is defined by separating out the number of times the rocks have been "tectonically" reactivated.

With each tectonic break a new set of infill can be placed in a relative time sequence.

In the above example the eagle eyed observer can interpret a tectonic break by carefully inspecting the sequence in the region of the small side vein (bottom left). Several layers of silica seem to have been deposited before this vein opened, although even at this scale the situation is not totally clear. The gross change in style in the central zone with the introduction of cassiterite is also extremely suspicious, even though no crosscutting relationship can be observed.

The sudden change from coarse-grained crystal format to fine-grained layer format is also suspicious denoting a change from isolated nucleation to multiple nucleation sites.

Obviously the total answer cannot be accurately deduced from this specimen, and more would have to be collected from the locality. However, there is enough to suspect at least three stages with several changes of chemical and/or physical parameters. Cassiterite is a relatively high temperature mineral (say plus 200 °C).

There is also a minute crosscutting silica vein (bottom left) which traverses the side layers of the small vein and could link to the centre quartz? The cavity development across the centre seems to be a late leaching effect? One late fine black layer seems to be absent from the top sequence!

0
1 cm

◻ PLATE 2

SEQUENTIAL INFILL – EPITHERMAL STYLE
Pajingo gold mine, Charters Towers region, Queensland, Australia (Scott Lode).

This plate has been selected to illustrate the problems of sequential infill in systems that are characterised by an excessive number of stages. It also illustrates the characteristic finely crustiform/colloform layering which occurs in many gold-silver epithermal systems.

Epithermal systems occurring in high level volcanic associated environments are commonly subjected to multiple and/or continuous fluid flow, which can be interrupted by repeated tectonic activity, including violent boiling and explosive events. Periodic boiling during a continuous fluid flow can result in rapid semi-rhythmic deposition to produce finely layered crustiform and colloform textures. Brecciation styles include explosive, and fault related products.

Faced with this situation there is often little point in trying to accurately define the number of individual phases in the conventional manner. It could easily exceed fifty and vary throughout the system, as different regions boil and/or fracture at different times.

The specimen graphically illustrates the problem and it becomes almost futile trying to sort out the brecciation and infill sequencing. Despite the very clear photograph the observer will quickly run into problems trying to piece the story together. Obviously it depends upon how desperately the total sequence is required. The beginner can rest assured that in 99.9 % of cases a rationalisation is adopted, with such statements as "multistage deposition" being employed! Strangely enough despite the multistage nature of many of these deposits the actual gold is typically restricted to a very narrow (or several very narrow) time bands. This is rarely acknowledged in scientific accounts which often "fudge" or ignore the paragenesis. The wall-rock fragments in the above plate are "fine-grained" silicified "volcanics". The Pajingo deposit consists of a vein system, which is bonanza style in that some sections such as the Scott Lode, reach three figure gold values. High grade zones are often dark coloured as silica layers associate with fine-grained sulphides.

0
1 cm

PLATE 3

SEQUENTIAL INFILL. STRUCTURAL SUPERIMPOSITION – CROSSCUTTING VEIN. Telfer gold mine, Telfer, Western Australia, Australia (M 10 Reef).

This plate has been included to illustrate a more complex example of sequential infill accompanied by fracturing. It covers the criteria of mineral superimposition and structural superimposition.

The left hand side depicts a complete vein section. The host rocks are carbonate units which are just visible as dark slivers on the top contact.

The basal zone commences with grey / white quartz and progresses to a mixture of quartz (grey/white) chalcopyrite (brass yellow) and pyrite (pale yellow). This passes upwards into a finer-grained zone of quartz, pyrite with minor chalcopyrite. The complex upper section consists of prominent bands of granular pyrite, white calcite, grey materials (dolomite and silica) and a network layer of pyrite/chalcopyrite which is similar in texture to the region just below the prominent calcite zone.

The enlargements show the details of the upper and lower regions of the vein.

The lower section (bottom right) is composed of grey/white quartz which is clearly some form of infill. Hexagonal crystal outlines, partially hexagonal outlines, corroded looking grains are all evident, and outlined by excellent triangular textures of chalcopyrite/pyrite infilling around the hazy quartz crystals. Pyrite crystal shapes suggest that it nucleates prior to chalcopyrite. The quartz grains are also characterised by vague dark rims which suggest some form of alteration (recrystallisation of silica?) relating to sulphide deposition. It is also apparent that the sulphide component cuts through the quartz crystals in places (bottom right). A brown mineral is also present as infill?

The texture is difficult to interpret as many of the quartz crystals appear to nucleate from no obvious substrate. They are clearly earlier than the sulphides, and from their appearance seem to be either fragments exhibiting mild corrosion, or a network of crystals growing in a random lattice format. The latter would provide the space required to produce the course-grained sulphide infill. The crosscutting sulphide relationship suggests mild tectonic disturbance either prior to, or accompanying, the fluid responsible for sulphide deposition. Has the fluid flow/pressure disrupted and mildly corroded preexisting silica crystals prior to sulphide deposition?

The interstitial sulphide becomes pyrite-dominant upwards, and both the grain size of quartz and interstitial void space decrease in size. This is a consistent feature of the saddle reef vein style of the M10 Reef at Telfer. This obviously requires some explanation. The texture suggests more rapid nucleation. The entire upper section of the specimen presents many textural problems. The interstitial pyrite-quartz (chalcopyrite) textures are present but are disrupted (overprinted) by prominent, vaguely vein-like layers of finer-grained duller pyrite, and also contain significant amounts of carbonate (white).

In detail the finer pyrite layers are composed of pyrite crystals of extremely variable grain size, and it is difficult to interpret whether they represent an infilling or an alteration component. The writer would suspect rapid nucleation with infill being dominant, although there is an impression of shear style in the fabric suggesting alteration pyrite could also be present. The problem cannot be resolved by eye.

The carbonate (white) component also appears to be vein-like infill where it traverses the width of the rock, and small tension style gashes are particularly prominent above the upper fine pyrite layer (top left) which could well relate to the original main horizontal fracturing. The white wisps, in and around the fine pyrite layers, again suggest shearing fabric associated with this zone. The carbonate thus seems to postdate the bulk of the main pyrite-chalcopyrite and silica stages, but predate the finer pyrite layers.

It is conceded that not everyone might agree on the precise details, but the general pattern of open space infill grading from coarse to finer moving upwards, with later stages of finer and/or shear? related components occurring within the top quarter, is present throughout all the underground exposures of this vein. Scheelite is also recorded at some points within the M10 Reef.

The Telfer geometry of stacked crescent, saddle reef shaped structures related to fold outlines requires interpretation within the above textural context. Early high fluid pressures have probably played a major early role in opening up the appropriate structural zones, with later shear? related fabric becoming focused in the upper portion. It is not the intent of this description to speculate further, but simply to illustrate that the observational process of understanding overprinting from a combination of infill, alteration, and structural perspectives opens up a wide realm of crucial genetic thinking.

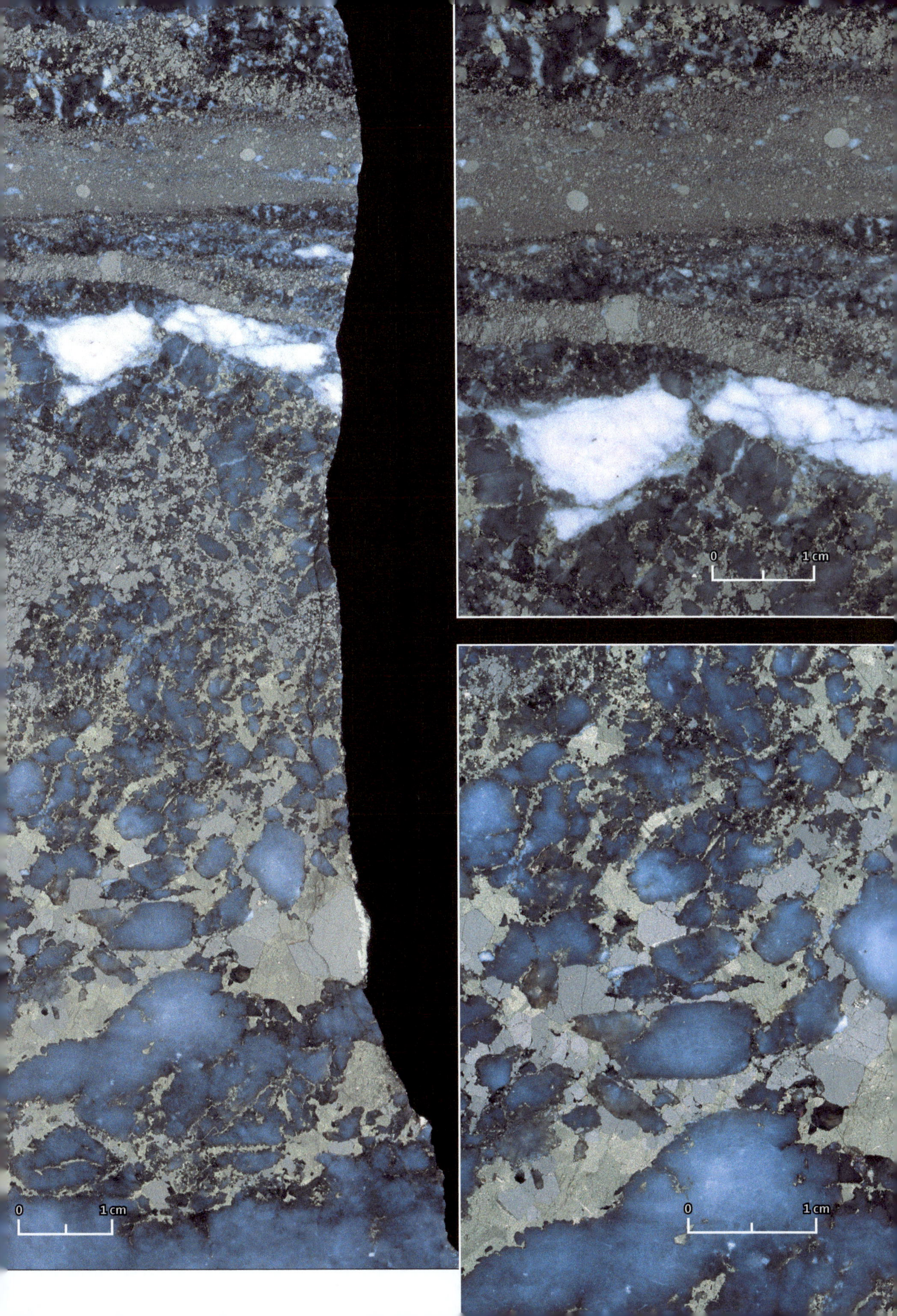
0
1 cm
0
1 cm
0
1 cm

5 First Order Criteria – Structural Superimposition

Crosscutting Veins – Stockwork and Other Vein Styles

Textures and Problems. Plates 4–11

Crosscutting veins are the most obvious of overprinting textures and usually present very few problems. Most systems consist of a relatively low number of stages (4–6) which are distinctly different and can be easily differentiated. However, there are some systems which by their very nature involve multiple stages of fracturing and deposition. The fracturing event may be closely allied to the depositional event. A good example would be epithermal systems, where repeated boiling over a short time span may result in 10–100 almost identical individual stages expressed as complex overprinting stockworks or multilayered colloform veins. Similar situations occur within some porphyry copper and porphyry molybdenum stockworks, and in fibre style crack-seal veins.

In some cases a complete sequencing is not only impracticable but impossible. Rationalisation is the usual approach, and typically large portions of sequence become arbitrarily lumped into one complex stage. This approach should not be used as an excuse for not really looking. Many porphyry stockworks are well worth pulling apart, with the complex sequence resulting from overprinting of a series of porphyries, rather than a single body. The recognition of nested porphyries each with its own system is a valuable piece of information.

Despite the apparent simplicity of the crosscutting criterion, it is quite surprising just how deceptive crosscutting relationships can be. The writer could recount hundreds of occasions where a casual visual conclusion had to be rapidly corrected when a hand lens was utilised. Quartz veins are particularly skilled at the art of deception!

The problems here mostly relate to the eye filling-in the crosscutting picture at the broad scale and not recording finer details. Typical eye deceivers are:

- An abrupt narrowing of a crosscutting silica vein as it cross cuts another silica vein. A prominent alteration halo around a quartz vein may disappear with the change in rock type as it crosses an earlier silica vein. This easily creates a false visual impression of being cut off. (See also ***Plate 9***).
- Late fracture reactivation commonly propagates across preexisting veins. A new fracture may propagate across a quartz vein, and link into earlier veins on either side. The net effect can easily mislead observers.

The following plates include examples of stockwork situations which illustrate the need for care.

◘ PLATE 4

STRUCTURAL SUPERIMPOSITION – VEINS
Kidston gold mine, Einasleigh region, Queensland, Australia.

This plate has been selected to illustrate that crosscutting relationships are not always as simple as might be expected, especially in systems dominated by silica. It also shows some typical laminated style quartz veins, which are very characteristic of porphyry molybdenum systems.

The relationship between the quartz veins within this plate is quite a mind puzzle. The central vein seems to cut everything and is composed of fairly uniform dark grey silica. There is little real evidence to delineate the proportions of infill versus alteration, although the relatively good matching of the walls suggest infill predominates. Against this it seems that some selective alteration has occurred of the vein intersection at the top right. A very close inspection reveals a very small darker silica vein within the central vein which is visible in the upper region of the plate and appears to contain wafer thin slivers of the host vein?

The central vein seems to occupy a fault zone, as none of the veins on either side can be confidently matched. The well-laminated (light and dark layered) vein (centre-left) appears to be symmetrical sequential infill and the dark spots are probably molybdenite. The sharp matching edges suggest little if any alteration. None of the other veins convincingly match this infill sequence.

The vein to the top right has some elements of the above sequence (especially to the extreme right), but seems to be cut by another generation of silica at a shallow angle. This is either cut off, or is running up the side of the main central vein. The minute link vein just below this intersecting vein set graphically illustrates the problem of decision making concerning which stage to group it with.

This account could continue describing all the other small veins within the stockwork. However, it is apparent that precise delineation of the number of stages is rather difficult. It is also apparent that all of the stages look very similar in general, and the system would be interpreted as a continuous fluid flow intimately associated with multiple periods of brittle fracture. Fluid pressure is difficult to estimate but seems enough to hold veins apart? The fine-grained nature of the laminated silica and molybdenite suggest rapid deposition with multiple nucleation sites. These observations trigger a series of questions. What is the cause of the repeated brittle fracture which appears to be part of a continued tectonic event within a relatively short period of time? What is the cause of rapid and/or fine-grained nucleation? Why is alteration so limited? Why is the lamination and probably molybdenite deposition not present in the central vein system? Why does the stockwork exhibit a vague vein parallelism within the darker silica stages? Why is this style of veining so characteristic of porphyry-related molybdenum deposits?

Molybdenite is generally regarded as a high temperature, magmatic-related mineral. The rock is actually a fragment within the main Kidston breccia pipe.

0
1 cm

◘ PLATE 5

STRUCTURAL SUPERIMPOSITION – VEINS. SUSPICION AROUSING – ASYMMETRIC INFILL

Watsons (lead, zinc) mine, Century region, Queensland, Australia.

This plate has been included to illustrate the difficulties involved in observing crosscutting relationships where the mineralogy of successive phases is similar. This is especially true of silica veins.

The plate illustrates several stages of silica veining, which consist of infill with very little alteration. Clean cut off sets have occurred, involving very little gouge effects. The cracks have opened under fluid pressure, such that they have reasonably well matching walls. This criterion is often used to identify infill from alteration. A few slivers of wall rock are included within some of the smaller veins, and in some cases seem fixed as "floating" fragments (central region). No one seems to have offered convincing explanation for "floating" fragments in silica veins which occur commonly at much larger scales. In many cases they are of high SG and should sink! Are they held apart by fluid pressure long enough for silica to precipitate commonly as comb quartz? Does the silica go through a gel phase at some time?

The most obvious stage is represented by the widest vertical veins, with the pale cream material consistently on the left hand side. The pale cream material (carbonate?) immediately arouses suspicion as it is blatantly asymmetric. Is this a later vein creeping up the side, or is the silica vein opening on one side of a preexisting carbonate vein? There is very little evidence to support either hypothesis. It is noted that the carbonate vein is actually at different distances away from the dark wall rock margin at different points, and that at some points (top) triangular textured cream infill seems to fill in between the quartz crystals. This favours a later origin for the carbonate. Against this these are a few points where the carbonate vein is discontinuous, with short patches of silica between the gaps. On balance the author favours the first hypothesis.

The obvious offsets of the main vein, contain a second silica stage which is slightly darker. A third even darker stage of silica angles across at about seventy degrees (bottom middle – to right middle). This also contains late cream carbonate infill and fragment(s) of the second stage silica.

Readers are invited to note the scale and reflect upon how difficult this would be to observe under normal conditions.

0
1 cm

◘ PLATE 6

STRUCTURAL SUPERIMPOSITION – VEINS, STOCKWORK.
Grasberg (copper, gold) mine, Ertsberg district, Papua, Indonesia.

This plate has been included to illustrate the complex detail that can occur within a porphyry copper stockwork. It is definitely not for the faint hearted. The stockwork is so intense that there is little recognisable rock texture remaining.

The latest stage recognisable is the green malachite resulting from surface oxidation. A copper-rich fluid has penetrated a small crack along the contact of the quartz vein (bottom left) resulting in micro-infill along the crack, and minor alteration of pink feldspars.

The obvious silica vein (bottom right) has been overprinted by a crackle style of brecciation, channelling fluids throughout the rock and precipitating pyrite (pale yellow) and chalcopyrite (brass yellow) as micro-infill, in a complex spider web stockwork. It seems that pyrite precipitation precedes chalcopyrite, although this is questionable. No alteration is evident. Note that at some points the sulphides occupy a crack in the centre of a quartz vein. This situation is common and if seen in isolation can easily misguide the observer.

The quartz vein(s) appear to have no alteration component and are composed of sugary grey and white infill. Their relatively high width suggests high pressure fluid control?

The remainder of the plate (top left) is occupied by very small black veins of biotite occurring as infill. Less obvious are the grey/black small veins and rounded grains of magnetite, and the remnants of a fine-grained porphyritic igneous rock (diorite?). Although timing evidence is a little contradictory, the majority of the black biotite veins cut the magnetite. This tiny wispy vein style is very typical of secondary biotite in porphyry coppers, and although usually referred to as potassic alteration they are actually potassic infill! The biotite alteration occurs within the ferromagnesian minerals of the host rock. The square/oblong black clots in the top left corner, are probably hornblende converted to very fine-grained aggregates of biotite. The conversion blurs the edges of originally sharp bordered phenocrysts. It is very probable that the pink feldspars are also alteration products of plagioclase, and are now changed to potassium feldspar. (The biotite/potassium feldspar assemblage constitutes the classic potassic alteration of porphyry copper systems. In the more mafic style porphyries, biotite prevails in relation to potassium feldspar).

The magnetite thus represents the earliest stage of stockwork brecciation/alteration. Small veins are visible, and the rounded magnetite grains are alteration of the host rock feldspars.

The general sequence of stages – magnetite-biotite silica-sulphides represents a declining salinity and temperature sequence (600 °C to 350 °C?) The intensity of stockworking implies repeated fracturing (semi-random shattering) by a process of repeated high energy release. (Gas release? Explosive? Hydraulic overpressuring?). The exact mechanism for multiple fracturing in porphyry copper systems is unknown. It is quite common for two stages of identical character to overprint each other, suggesting a strong link between fracturing and fluid release.

0
1 cm

◘ PLATE 7

STRUCTURAL SUPERIMPOSITION – VEINS, STOCKWORKS.
Pea Ridge, Missouri, United States of America.
(Specimen provided by R. E. Myers).

This plate has been selected to again illustrate the complexity and difficulties encountered even within well-defined stockwork systems. The real question here is just how many stages of brecciation and infill are involved?

Disregarding the minute vertical cracks which look like very late veins, the latest stage is clearly represented by the system of angled (right to left) veins which contain a variable mineralogy including quartz (white-grey with vugs at centre right) fine-grained grey material (magnetite, sulphides) and equally fine-grained greenish areas (epidote?). The stage is infill-dominated, and it is difficult to discern any alteration effects. There is a very good example (centre left) of an abrupt right angle turn where the stage swings into a preexisting vein orientation.

It is a little difficult to ascertain the next stage back, but it seems to be a dark-to mottled grey (silica-sulphide/magnetite) vein system, best seen in the bottom left hand corner where it clearly cuts both the pink and green material, and is in turn cut by the grey material described above. This stage is present as small veins (bottom left) and as minute fracture networks (E-W, across mid slide). Alteration is variable. The pink wall rock seems to suffer very little, whereas the green (epidote) becomes darker. The darker alteration (silica?) is particularly noticeable in the E-W anastomosing fracture network (centre).

The next stage is probably the clearest, being represented by the very visual green (epidote) in both breccia and small vein format. There are numerous examples of well-defined infill (green epidote, with darker green quartz) which have nucleated on vein walls and around fragments. It is possible that there is some whiter quartz in the centre of larger veins (bottom left). Alteration associated with this infill seems negligible. Fragments are predominately pink and dark green/black and the breccia appears to be fault style, probably associated with high pressure fluid introduction (hydraulic jacking).

The early history of the rock is best seen in the top third of the plate. The epidote stage clearly overprints a dark black stage (silica?) which occurs as breccia fragments and as asymmetric thin edges on veins which have been reopened by the epidote hydrofracking.

Also present in the top left hand quadrant are minute pink infill veins (feldspar?) which cut a dull grey (silica-magnetite) background crackle-mesh network breccia system. This system appears to have a considerable dark silica alteration component. From this examination it would seem that the rock has been broken up and subjected to hydrothermal fluid systems on at least six occasions. No doubt the reader would question some of the above, as it is clearly a complex overprinting situation requiring very good observation and extreme care. The specimen has been included to make exactly this point. Very commonly all is not obvious, and more specimens, together with controlled petrology are required. The original rock is a little enigmatic. The pink crystals are presumably feldspar, the white blebs are pyrite, within a very fine-grained matrix. A species of pophyritic volcanic or intrusive rock is suggested.

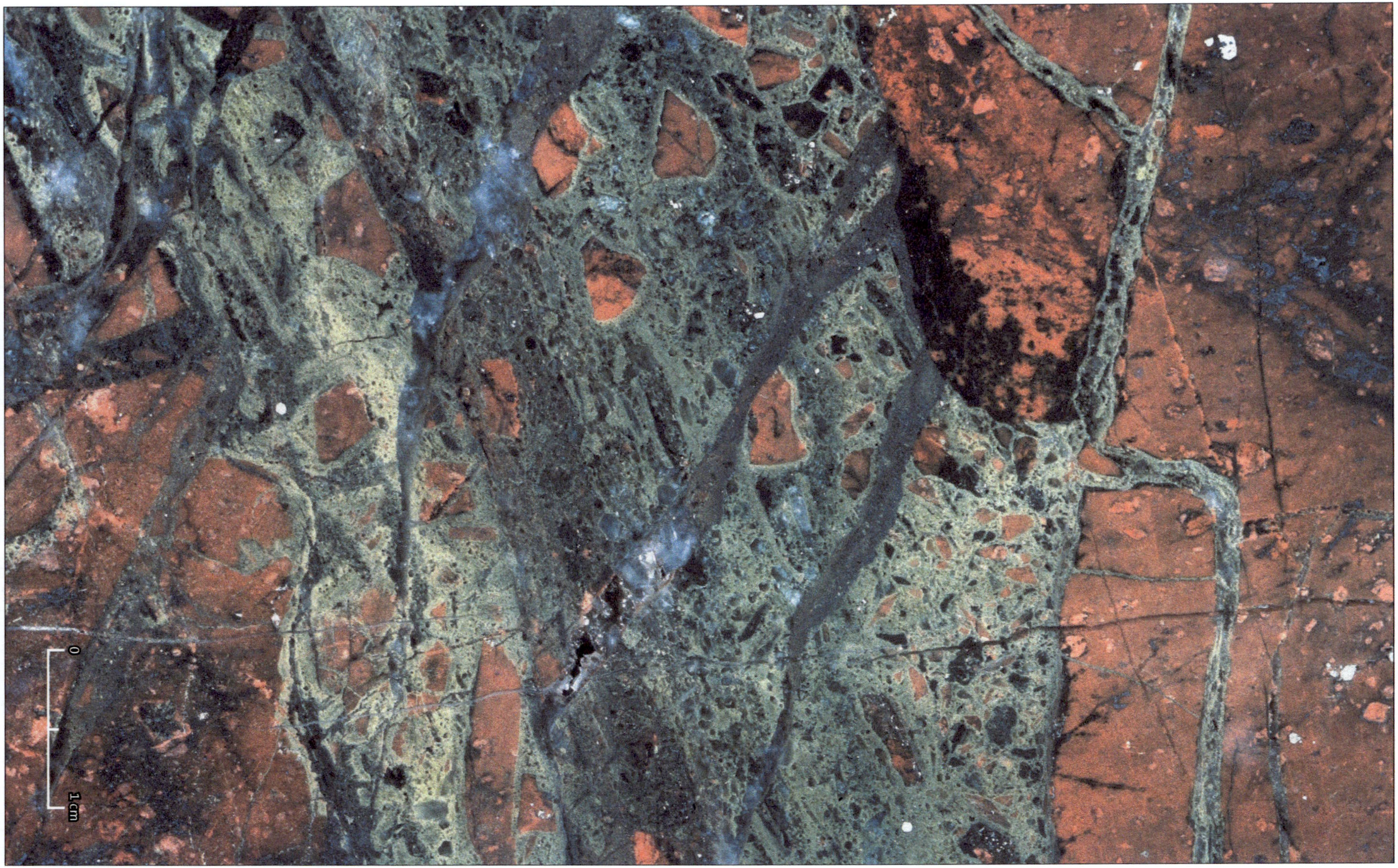
0
1 cm

◻ PLATE 8

STRUCTURAL SUPERIMPOSITION – VEINS STOCKWORKS.
Red Dome (copper, gold) mine, Chillagoe, Queensland, Australia.

This plate has been included as a good example of a multiple overprinting stockwork. It also illustrates that even with clearly visible components, timing relationships and alteration/infill distinction requires very careful observation/consideration.

The last stage within this rock is represented by the prominent dark, vertical, central vein of bornite. The same stage is visible at many other points as impersistent lines (small veins) of dark spots, and vague alignments of diffuse vein-like spots/blotches. The latter are prominent within a vertical yellow band (garnet) on the right hand side. This style of texture is common in skarns, and all of the dark material marks subtle channelways. The identification of infill versus alteration is impossible, although it is clear that most of the spots are alteration. Some infill may be present in the main vein?

The preceding stage is represent by the narrow veins (grey quartz) which run E-W across the plate. These are infill (matching walls), and best represented at the top of the plate. Close inspection of this region reveals that despite initial impressions the quartz vein does not actually cut through the bornite. Similar relationships are more obvious towards the base of the plate.

Both the bornite and quartz veins overprint the rather vague diffuse yellow garnet veins (right hand vertical, and top left corner). Sight variations in texture are visible, but the infill/alteration distinction is not clear (probably mostly alteration). The yellow garnet occurs as alteration of red garnet (bottom-centre).

At this point life becomes difficult! The obvious remaining materials are garnet (red-centre vein/layer) and white (wollastonite). Wollastonite clearly cuts over the red garnet (bottom left quadrant). Careful inspection of the general white colour of the rock reveals grey zones, which suggest that mass wollastonite alteration of an unknown host rock (siliceous carbonate??) has occurred.

Infill/alteration distinction of the wollastonite stage is impossible, although obviously alteration predominates.

Even closer observation reveals that the red garnet is cut by impersistent dark green vein structures (pyroxene). These are best seen in the region of the red garnet/wollastonite crosscutting vein relationship (bottom left quadrant), and the wollastonite also cuts the dark green pyroxene veins?

The reader may well be reaching a point of scepticism by this stage, but this sequence can actually be observed from other rocks in the mine area, and is records two overprinting stages of skarn development.

The short form paragenesis becomes:

1. Red garnet veins (alteration predominant, textural variation suggests some infill?).
2. Dark green pyroxene veins (alteration predominant? Not really visible).
3. Wollastonite (mostly as alteration, minor infill).
4. Yellow green garnet veins (mostly as alteration).
5. Quartz veins (infill).
6. Bornite veins (mostly as alteration, possibly minor infill?).

Obviously more specimens are required to achieve more information on the early stages, and wall rock composition. The specimen is however a very good example for honing observational skills!

0
1 cm

◻ PLATE 9

STRUCTURAL SUPERIMPOSITION – VEINS. SUSPICION AROUSING – WANDERING CENTRAL INFILL
Jumna tin mine, Irvinebank, Queensland, Australia.

This specimen has been included to illustrate that overprinting vein structures can be considerably more complex to unravel than is popularly supposed.

It is always good technique to work from the latest event through to the earliest.

On this specimen the latest event seems to be a series of very inconspicuous tiny pale brown veins (siderite) which are best seen in the lower central region where they cut through the dark grey-white zone as well as both quartz veins and the black (quartz-tourmaline) veins. In this region they are infill-dominated and without alteration halos. However, this is a trap for the inexperienced as the same veins in the upper region produce irregular alteration fringes of orange (rusty? siderite) and pale creamy-orange materials (pristine siderite?). This appears to be at the expense of greenish sericite in the host rock. The latter is some form of feldspathic sandstone, in which the feldspars are sericitised (top zone).

Thus the alteration associated with this stage is dependent upon the rock type. Within this context it is probable that the conspicuous orange/cream central "vein" is in fact part of this assemblage, representing alteration associated with an E-W fracture crossing a zone of relatively unaltered host. A very good example of a false central channel is present within the lower narrow black vein. Close inspection reveals a minute central pale brown vein of the siderite stage wandering along the horizontal dark tourmaline infill, but connecting to the vertical system.

The next set of veins are the dark horizontal set which consist of dark infill (tourmaline) and variable amounts of grey/white (silica) and dark (tourmaline) alteration. The latter probably alters an original feldspathic component in the rock and may grade out into the green sericite mentioned above. It is obviously a strong alteration in relation to the small infill channels.

This stage again illustrates the need for very careful inspection of seemingly crosscutting veins, as there are many examples where the tourmaline veins cut earlier quartz veins, but turn to quartz infill without alteration. This is common in many situations and can easily give the wrong impression to a superficial glance.

The earliest vein set is the white-grey quartz veins which contain dark (tourmaline) central zones and coarsely crystalline comb quartz. The veins display no alteration and their relative width suggests a high fluid pressure. There is also a hint of a brown mineral (cassiterite) at some points.

A paragenetic/structural control chart is shown below. This format is very useful for recording data.

Stage (Name)	Infill	Alteration (Rock type altered)	Structural Style	Comment
Siderite	Siderite, mm scale	Siderite (sericitised? wall rock) Nil (Silica-tourmaline rocks) Nil (Quartz veins)	Microvein network mm scale	Small scale
Tourmaline	Tourmaline mm scale	Quartz tourmaline (sericitised? wall rock) Originally feldspathic sandstone)	Subparallel system of small discontinuous microveins	Alteration extensive
Quartz tourmaline	Quartz tourmaline cassiterite ? cm scale	Nil	Subparallel veinlets around 1 cm wide	Fluid pressure high? Comb quartz

0
1 cm

PLATE 10

STRUCTURAL SUPERIMPOSITION – VEINS, PARALLEL REACTIVATION.
Unknown tin mine, Kangaroo Hills tinfield, Queensland.

This specimen has been selected to illustrate the difficulty in establishing overprinting relationships in situations where infill is not accompanied by significant alteration. It also illustrates a much overlooked situation in that feldspathic alterations are commonly accompanied by feldspathic infill.

The wall rock (bottom left) is composed of pale (pink/white) fine-grained granite, composed mostly of feldspar (pink/white) and quartz (grey). This has been altered (bottom right) to a texturally retentive rock with the original feldspars converting to white albite. The infill component for this is represented by the coarser-grained white albite associated with black spots of cassiterite and grey quartz. Careful observation will reveal small crystals of oblong albite projecting into the small "triangular" cavities containing cassiterite crystals.

Subsequent fracturing along the central and marginal region of the vein, has provided access for fluid depositing coarse cassiterite in an anastomosing vein mesh. This is not accompanied by any obvious alteration and possibly occurs in close time sequence to the cassiterite associated with the albite infill?

The major quartz stage (grey) appears to be all infill (top and middle) occurring as crystals, although it is possible that some of the paler albite rich slivers (top left) might be partially silicified. The timing of the quartz stage is a little difficult to discern as it appears to have cassiterite (dark) as a partial lining at several points. However on balance it seems more probable that a high pressure fluid has exploited previously earlier cassiterite veins and simply forced (hydrofractured) them open.

The prominent orange layer (lower middle) is iron staining related to ground water percolation along the vein margin.

0
1 cm

PLATE 11

STRUCTURAL SUPERIMPOSITION – VEIN, BRECCIA
Mt Isa (lead, zinc, copper) mine, Mt Isa, Queensland, Australia.

This specimen has been selected to illustrate a very clear cut case of overprinting relationships in breccia format.

The most striking feature is the comprehensively shattered, crackled white material (dolomite) which is associated with a network of narrow veins containing chalcopyrite (yellow) and dark material(?). The vein network links to more substantive blotches dominated by chalcopyrite and minor pyrite (paler yellow).

It is very difficult to determine whether or not the chalcopyrite represents infill or alteration. This is a common problem with chalcopyrite which very rarely forms clearly recognisable crystals. The substrate dolomite (white) also gives very little help in terms of providing clearly recognisable crystals projecting into potential chalcopyrite-filled vugs. Some of the ragged edged dolomite and small fragments? (mid to lower right) suggest that dolomite may have been dissolved prior to chalcopyrite deposition or simply altered to chalcopyrite? However, without further data it is impossible to be certain of the proportions of infill versus alteration concerning chalcopyrite. The writer would suspect both are present.

The fracture system controlling chalcopyrite also cuts the dark (siliceous) fragments, and although not totally clear may link with some dirty grey dolomite veinlets (bottom right) associated with grey spotting (alteration) within the fragment.

The relationship between the dark siliceous fragments and the white dolomite is especially interesting. Taken at face value there is almost no evidence to suggest that the fragments have been altered. They have sharp edges, and there are several examples of fragments which are veined by dolomite. In these cases the sharp walls can be matched and/or partially matched, leaving the overall impression of high pressure fluid jacking with subsequent coarse-grained dolomite infill. The current concepts at Mt Isa which regard most of this dolomite as alteration (replacement) certainly merit re-examination.

The siliceous nature of the fragments is also very suggestive of alteration prior to fragmentation.

All of this suggests a three stage paragenesis of:

- Early dark silica alteration (channelways uncertain?)
- Major fracturing and tectonic brecciation involving a high pressure fluid which precipitates dolomite (mostly infill).
- Major fracturing and brecciation/crackling linked to the introduction of the copper-rich fluids. (proportions of infill v alteration uncertain).

This specimen is typical of much of the copper orebody. Obviously some good petrographic observation is required to sharpen this perception.

0
1 cm

6 First Order Criteria – Structural Superimposition

Breccia (Fragments of Early Stage Mineralisation Contained Within Later Stages)

Textures and Problems Plates 12–15

Recognition of overprinting utilising this criterion would at first reading seem very straight forward, and there are many breccias where relationships are extremely well-defined. However, anybody aquainted with the breccia world, will readily testify that coping with mineralised breccias is frequently one of the most difficult textural operations in economic geology. Indeed there are numerous examples where brecciation was not even recognised during many years of production.

Brecciation produces a mix of void spaces, large fragments, and finer fragmental (matrix) materials. The variation in scale, proportions, and distribution of these components is enormous. The highly broken zone is usually flanked by regions of shattered and mildly jostled rock. The highly permeable result forms a perfect focus for migrating ore fluids. Hydrothermal activity results in alteration of both fragments and matrix with precipitation occurring within the voids. Fluid flow through large bodies of breccias is usually very irregular, producing very erratic distribution of alteration/infill.

Breccia bodies (especially fault-related breccias) are commonly sites of reactivation. Rebrecciation and new fluid flow can produce complex overprinting textures. The new fluid may precipitate in already partially filled voids, or occupy the new voids. Similarly alteration may alter preexisting alteration (or infill) or attack areas of original unaffected breccia. The process may be repeated several times.

Common problems include:

(a) Recognition of altered fragments
(b) Distinction between highly altered fragments and similarly altered matrix.
(c) Distinction of infill textures, and their relationship to the above.
(d) Infill/alteration pairing. Within multibrecciated systems which involve extensive alteration, it can become extremely difficult to decipher which alteration belongs to which infill.

In complex breccia overprinting situations, the observer is strongly advised to immediately cease activities and move towards the edge zones. The "crackle" zone at the periphery, is highly likely to contain a mass of minor overprinting veins. Infill and alteration pairing becomes much easier utilising a consistent wall rock, crosscutting textures are more obvious, and the information can then be applied to the main body. Unraveling overprinting within complex multiply reactivated breccia zones requires good technique, good observation, and patience.

PLATE 12

STRUCTURAL SUPERIMPOSITON – BRECCIA
Disraeli gold mine, Rishton district, Queensland, Australia.
Specminen provided by G. W. Clarke.

This specimen has been selected to illustrate many of the problems associated with overprinting in breccia mode. These are three phases of hydrothermal fluid introduction involved, and three types of fracturing/brecciation. Working with brecciation is frequently much more difficult than is generally conceived, and many breccias are overlooked. Readers are invited to test their observational skills on the opposite plate, before reading the detail below.

The last stage is easily the most obvious, consisting of calcite (white) and minor silica (grey) infill without alteration, occupying a crackle style brecciation. (This stage is actually overprinted by another crackle style fracturing which is unmineralised and possibly caused by blasting activity.)

The middle stage is best observed by tracking the pale yellow-buff materials, which are varying combinations of sericite and pyrite. These actually all link together in a complex network at both a coarse and fine scale of fracturing. The differentiation between infill and alteration within this stage is not really possible. Intuitively it seems that pyrite is involved as infill, but it also probably occurs as an alteration product. Sericite is clearly involved, largely as alteration.

The mode of brecciation is a stockwork of brittle fracture which seems to have involved the production of very small fragments of wall rock (fragments and gouge associated with a crushing/shearing movement). This matrix forms the main focus for alteration (top third of plate) and fluid flow.

When this breccia is eliminated the rest of the rock remains, and is composed of well-defined dark fragments and grey silica. The fragments are "granite", and are well-separated with many being well-rounded. One "granite" fragment has an attached silica edge and other smaller silica fragments are present. The matrix of the early breccia is very difficult to discern through the overprinting sericitic alteration. It seems siliceous? and possibly finely fragmental. (The fragment at the top right seems to be different and possibly silica-altered). All of this suggests that the early breccia may well be of the hydrothermal intrusive (milled) variety. It is possible that some silicification (quartz veining) occurred before this brecciation. If this were so, it would imply another phase preceeding the interpreted silicified hydrothermal intrusive breccia. Obviously a wise observer would move and try to find less overprinted material of the early stages. However, there is enough within the plate to lead the observer to suspect an intrusive breccia pipe environment?

Selective assaying is required to track the gold. The middle pyrite-rich stage is the most probable candidate.

0
1 cm

PLATE 13

STRUCTURAL SUPERIMPOSITION (AND MINERAL SUPERIMPOSITION) – BRECCIA Ravenswood gold mining district, Queensland, Australia (Area 2 vein Carpentaria Gold).

This specimen has been included as an example of complex infill overprinting and brecciation which reflects a long history of deposition within a vein structure.

With situations of this style it is always very difficult to know where to start. The working backwards principle (from the latest to earliest) is not always appropriate. It is often more effective to settle on one small segment that is clear, and then select another. This can often be done by working from clear crystal shapes to see the next mineral to be deposited, or alternatively from a layer which is coating either crystals or broken material.

A most striking feature is the strongly brecciated material (lower left hand edge). This is predominately finely comminuted sphalerite (black) and arsenopyrite (silver) within a black (silica?) matrix. The fragments are sharply angular to well-rounded, and give a very good idea of how the two different sulphides break up. The breccia breaks through preexisting sulphides, and fragments have clearly been forcefully transported. Some style of explosive/intrusive brecciation (milling) can be inferred as a late-stage event.

In the left hand top corner a continuous rim of arsenopyrite (silver), some with good crystal form coats a large sphalerite crystal (dark). The latter seems to exhibit a crystal shape, although it does look a little "ragged" along the margins. The arsenopyrite rim is directly overgrown by chalcopyrite (yellow) associated with some good crystals shapes of pyrite (pale yellow).

Direct sequencing is difficult here. Has the arsenopyrite been fractured to form isolated clumps? Which came first, chalcopyrite or pyrite? The author's preferred interpretation is that the arsenopyrite has been disrupted and that the pyrite is slightly earlier, in a pyrite-chalcopyrite paring.

The pale buff material, siderite? (right hand side) is quite clearly infill with good triangular textures, and equally clearly some of the earlier arsenopyrite/pyrite/chalcopyrite has been disrupted prior to siderite deposition. The 20/20 vision observer may just see a rare thin arsenopyrite layer coating some of the pyrite preceding siderite? deposition. Fragments of the pale buff siderite occur in the finely brecciated (black) material mentioned above (middle-right hand edge). Some lozenge shard-like cream crystals (calcite?) also appear in this region, and seem to coat the buff siderite, but predate? the violent brecciation. Their precise association is not clear.

Obviously a few more rocks might clear up some of the problems. However, the smorgasbord approach has partially unravelled the problem, which is one of vug filling with periods of selective fragmentation. The sequence appears to be sphalerite (possible fragmentation) arsenopyrite (probable fragmentation) pyrite-chalcopyrite (definite fragmentation) siderite-calcite (explosive fragmentation) dark silica. Much of the fragmentation is probably in response to high fluid pressures within vugs rather than direct tectonic brecciation. The more vigorous "explosive" event could be triggered by fluid boiling and streaming gas (CO_2?) discharge.

0
1 cm

PLATE 14

STRUCTURAL SUPERIMPOSITON (AND MINERAL SUPERIMPOSITION) – BRECCIA
Mistake tungsten mine, Emuford, Queensland, Australia.

This specimen has been included to illustrate the prime value of breccia in paragenetic sequencing. The working principles involved with breccia are firstly to determine the most recent event, and then to work progressively backwards within the remaining fragments. The process continues as different fragments become isolated. In complex brecciation it is sometimes more practical to just work in small segments (where things are clear) and then piece the results together.

The most recent event on this plate is the central white vein traversing diagonally from top left to bottom right. The picture is a little confused by the orange iron staining, but the vein consists of central white silica with a lining of grey-green-buff fluorite. At this point the situation becomes confusing as the two sides of the lining do not match very well. A thin zone (top left) of creamy fluorite runs along the top edge of the lower lining, and although difficult to trace seems to cut off a vein of pale grey-green-buff fluorite traversing diagonally from bottom left to top right. The net result is a degree of confusion as to whether there are one or two generations of the grey-green-buff fluorite (two are suspected).

Working backwards becomes difficult from here. The basal segment (base of plate) contains several clearly recognisable fragments (mottled buff-green) and numerous less obvious fragments (grey-green to grey-white). Good infill recognition is required at this point to recognise the small black crystals (wolframite) growing in association with dark blue-grey silica. A general (admittedly confused) sequence of dark blue-grey silica progressing to white silica is commonly picked out. The wolframite occurs in both, and despite the complexity seems to be a coherent infill package.

The fragments are obviously mixed, and clues to some of their origins are seen in the top right region. Here the mottled buff-green material is clearly surrounded by grey-green silica as a separate earlier breccia. This has been fragmented to produce the fragment mix associated with the wolframite-silica described above. A good example is seen of an early breccia fragment at mid-centre left. The mottled buff-green material is fine-grained granite wall rock. Several partially silicified examples can be located, and there is even a hint that this may have been broken by a white silica breccia generation prior to the grey-green silica!

The difficult breccia situation is even further extended with the realisation that a complete stage of pale emerald-green fluorite is present (top edge). This is cut by the late-stage veins, and appears to have deposited on top of the wolframite dark silica stage.

There is ample evidence on this plate to illustrate a minimum of four stages of brecciation and infill. The fine-grained granite has been fragmented and infilled by grey-green silica (stage I). This has in turn been fragmented and infilled by the wolframitegrey-white silica (stage II). The green fluorite represents another infill probably related to another brecciation (Stage III) and this has been cut by the late (grey-green fluorite) veins (stage VI).

Given that there could be two generations of the latter, and that both early silicification and sericitisation of the initial granite wall rocks are possible, there could be as many as seven stages involved! Obviously more samples would clarify some of the questions. It should be noted that the wolframite-silica stage is quite separate from the later fluorite stages. The specimen comes from a vein, and again illustrates the propensity of veins to constantly form focal points for fragmentation and new fluid flow. This complex breccia is nearly all composed of infill with very little alteration effects. If alteration had been a major component it is not hard to imagine how difficult it would be to unravel! The styles of brecciation represented are another aspect which needs consideration!

0
1 cm

PLATE 15

STRUCTURAL SUPERIMPOSITION – VEINS, BRECCIA

Unknown tin mine, Kangaroo Hills tinfield, north Queensland, Australia.
(Specimen provided by G. W. Clarke).

This plate has been included to illustrate the necessity for careful observation. It also serves to show some of the frustrations involved with low angle cross-cutting silica-rich veins and also to exhibit difficulty in observing breccia. The question to be answered is how many stages are present?

The plate contains multiple generations of quartz-rich veins some of which are involved with accompanying alteration. Vision is partially obscured by a yellow (iron-oxide) staining due to weathering, where surficial fluids have selectively channelled along vein margins and elsewhere. This is prominent in the mid-left of the plate, adjacent to a prominent quartz infill vein. This prominent silica vein has a Y shaped branching form and is distinguished by coarse-grained comb quartz infill. Within the comb quartz, dark triangular textures and cavities are visible. The dark mineral is unidentified, but seems to be prone to weathering. There is no clear-cut alteration associated with this stage. The vein width tends to suggest a fairly high fluid pressure which has pushed the walls apart. The lack of gouge or included fragments further suggest it was originally some kind of relatively simple cracking rather than a major displacive fault zone.

The lower branch of this comb quartz stage (base-middle) is cut off by a later similar, although slightly whiter silica vein in the bottom right hand corner. A close inspection of the comb stage silica vein, below the fork of the Y shape, reveals very narrow white silica veins within the comb quartz. These parallel the comb quartz vein, and can only be traced within it. Do they represent the crosscutting stage in the bottom right hand corner?

A third stage is strikingly represented by the narrow (vertical) band of dark material adjacent to the Y fork in the centre of the plate. This stage is an exercise in frustration. It appears to have a double black zone with a white centre at some points, but proves impossible to trace with any certainty through the various siliceous components.

The problems become even more acute when the various slivers of material surrounding the Y shaped comb quartz are inspected. For instance the bottom left hand sector is composed of vague fragments, some of which seem to be in the process of alteration (silicification) set in an anastomosing system of sugary granular slightly pink quartz. This breccia appears to be a very shattered feeder system, allowing easy ingress to an early-stage quartz infill/alteration system. Similar material is noted to the right of the vertical elongate dark central patch.

Even this dark material on close inspection is fragmentary and consists of dark green (chloritised) material containing reddish minerals (cassiterite?) which may link with the white silica to the immediate left which also contains the red/pink mineral. This all suggests another stage of silica (cassiterite) associated with chloritic alteration. The wall rocks are probably the grey/white fragmental materials (metagreywacke) seen as a vertical spindle shaped sliver, which just touches the middle-right hand side of the plate, and extends almost the full length of the plate.

At this point the observer will probably have collapsed in confusion, which was the reason for selecting this plate. Obviously things are not as easy as they seem, with overprinting veins/brecciation. More rocks would help!

The final!! interpretation is:

1. Fragmental metagreywacke wall rock (centre-right spindle)
2. Quartz-cassiterite infill with dark chlorite alteration (left and right of centre)
3. Quartz (sugary pink) infill and alteration (bottom left)
4. Narrow dark vein (centre) possibly part of 2 above?
5. Comb quartz vein (Y shaped)
6. Quartz vein (bottom right and possibly within the comb quartz vein)

0
1cm

7 Second Order Criteria; Suspicion Arousing

Textures and Problems Associated with Vein Re-Opening, Including-Overprinting Parallel Infill, Alteration-Infill Mismatching, Inconsistent Alteration Selvages and Alteration of Alteration

PLATES 16–22

The following section of illustrations has been selected to illustrate suspicion arousing or second order criteria.

The majority of these occur where preexisting channelway structures (usually veins) have been reactivated. The reactivation of planar structures naturally focuses the failure surfaces in a parallel, or semi-parallel, manner to the original fracture. Most commonly, failure occurs along vein margins, or directly up the central zone, and ranges from non-disruptive simple fracture propagation to disruptive brecciation. In most cases, the reactivation will also result in new fracture surfaces within the surrounding host rocks. These are especially concentrated around the fringe zones of the original fracture. Most of the new features are best approached by looking for, and collecting specimens whilst in the field, and this should be the first approach at normal visual scale. Thus at outcrop or drill core scale the first action should be to move away from the centre of mineralisation and seek isolated small veins within uniform host rocks. The objective is always to try and decide which infill assemblages belong to which alteration assemblages.

Fortunately infill assemblages are usually less affected by changes in host rock than alteration assemblages, and in the rare cases of multiple host rocks it quickly becomes apparent that variations in alteration do in fact belong to one mineralising stage and simply reflect the chemistry of different host rocks.

If the observer only has an individual slab/specimen of rock, and is unable to move there are many helpful suspicion arousing circumstances *(see page 105)*.

PLATE 16

SUSPICION AROUSING COMPLEX OVERPRINTING, PARALLEL REACTIVATION, STRUCTURAL SUPERIMPOSITION – VEINS. Ravenswood gold mining district, Queensland, Australia (SYC Pit? Carpentaria Gold).

This specimen has been included to illustrate the common situation of repeated re-opening, and infilling of the same fracture system. It also illustrates just how difficult this can be to decipher. The problem is compounded when each stage contains similar components.

The narrow vein system traverses a slightly "weathered" granitic rock, composed of quartz (grey-white) and feldspar (pink-orange iron-stained) with minor biotite? (rusty brown). The feldspars are slightly argillised and minor acid leaching has occurred via some of the crack systems. These traverse the vein horizontally and link to low-angle fractures (left hand side). No clear alteration halo borders the vein system. A close inspection of the vein indicates that it contains at least three separate components.

The upper-left vein section is noticeable, as it contains very well-developed white quartz crystals growing irregularly from the vein wall, succeeded by pyrite (pale yellow). This region is closely associated with small elongate brown components (mildly oxidised marcasite?). This assemblages can be traced along most of the left hand side of the vein, and then becomes vague, but re-appears in the bottom right hand edge. The bottom right hand occurrence is clearly cut off, by an assemblage of slightly greyer silica associated with arsenopyrite (silver). This can be traced along most of the right hand side of the vein, and is visible again to the right of the first stage assemblage at the top of the plate.

The arsenopyrite-grey quartz stage is in turn cut by another vein, which is less well-defined, and composed of dark brown (sphalerite) crystals, associated with pale buff (siderite?) material, and probably pale yellow pyrite and quartz.

The relationships between these stages are very confused and difficult to decipher. Indeed the author is by no means positive that the above interpretation is totally correct. This serves to illustrate just how effectively veins of similar composition can "hide" as they overprint along the same controlling structural weakness. Once again inspection of more examples is required to confirm initial suspicions. This example also illustrates the need for very careful observation. The vein is gold-bearing, and it is not certain which of the stages is gold-related!

0
1 cm

PLATE 17

SUSPICION AROUSING – COMPLEX OVERPRINTING, PARALLEL REACTIVATION. STRUCTURAL SUPERIMPOSITION – VEINS. Ravenswood gold mining district, Queensland, Australia (Buck Reef – SYC Pit, Carpentaria Gold).

This plate illustrates good second order suspicion arousing criteria which lead the observer to suspect two stages of mineralisation.

The first reaction to the rock is to note the sulphide-rich infill channel, and the attendant dark alteration zone. The host rock is tonalite, and the alteration is biotite. Biotite alteration is conventionally associated with minute veins in porphyry copper systems, but is actually quite widespread in many other magmatic-associated fluid systems (especially in mafic units associated with skarns) and can assume major proportions. The suspicion arousing circumstances relate to the thin veins, which extend at low angles into the host rock. These are composed of pyrite silica? (infill), but have no alteration halo. Given the intensity of the black biotite alteration this is an obvious anomaly.

Close inspection of the infill channel reveals two distinctly different materials. Coarse-grained pyrite crystals form one association with hints of black triangular textures (biotite?) occurring around the centre left region. A second association contains grey-white material (silica) with finer-grained elongate stringy pyrite. A small vein crosscutting the biotite alteration appears to link to this (bottom right).

The relationship between the two infilling stages is difficult to discern. Intuitively, the author suspects the biotite-related material to be early utilising the indirect criterion concerning temperature of deposition. Biotite alteration/infill implies high temperature (>400 °C). This plate is another good example of refracturing along a preexisting channelway, and again emphasizes the need for good understanding of the infill/alteration linking. Infill textures at this scale are commonly overlooked.

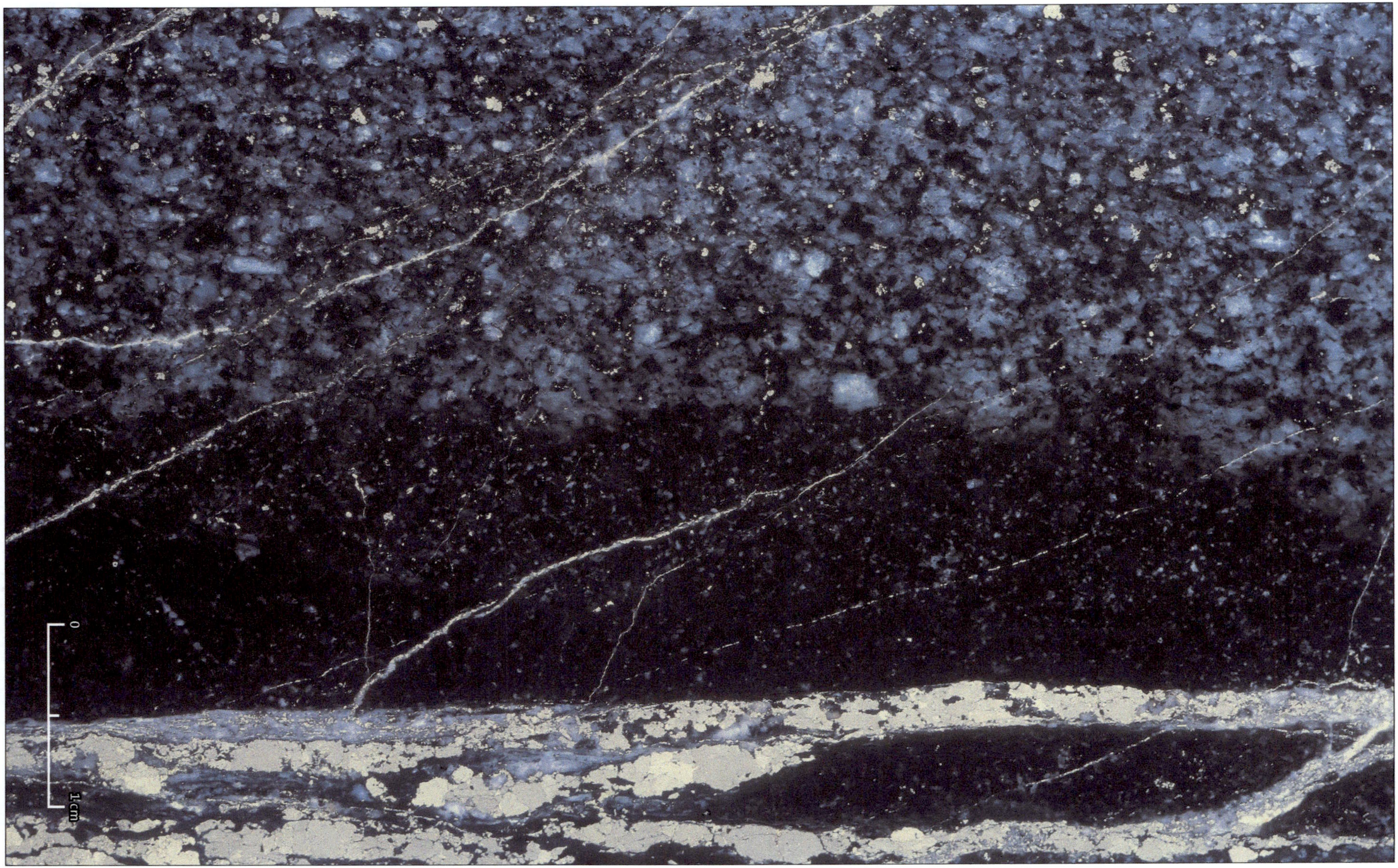
0
1 cm

◻ PLATE 18

SUSPICION AROUSING – COMPLEX OVERPRINTING PARALLEL REACTIVATION, ALTERATION MISMATCH. STRUCTURAL SUPERIMPOSITION – VEINS.
Ravenswood gold mining district, north Queensland, Australia (Nolans Pit, Carpentaria Gold).

This plate has been included for two reasons. Firstly it has many suspicion arousing overprinting criteria, and secondly it will make the point that it is unwise to try and resolve problems in the central regions of overprinting vein systems. Where faced with conflicts of mismatching alteration, or possible crosscutting indications, the observer is advised to move. Observation at the vein margins (crackle zones) or elsewhere, may quickly isolate individual stages and establish infill/alteration pairs.

The plate contains two sulphide veins which contain sphalerite (dark) arsenopyrite (silver) and minor pyrite (pale yellow). The sphalerite has a few silvery flashes due to cleavage reflectance. These appear to be infill minerals, and at first sight the arsenopyrite seems to line the walls of the channel. However, there are some suspicion arousing problems with this seemingly natural arrangement of high-temperature arsenopyrite overprinted by sphalerite. Firstly there is a disturbing lack of arsenopyrite on the side walls of both veins towards the top of the plate, and secondly there are several hints that arsenopyrite may be in tiny veins that cross the sphalerite. The sphalerite occurs in the typical coarse -grained rounded crystal form, and individual grains look to be broken open in the upper section of the right hand vein. There is not enough evidence to be certain, but from other specimens this is indeed the case. The arsenopyrite is later, and a fracture has utilized the sphalerite vein walls to channel in the arsenopyrite -precipitating fluid. The suspicion arousing criteria are giving the correct signal!

The situation regarding alteration pairing becomes even more suspicion arousing. The original rock is tonalite, and the darker rock units prominent on the bottom centre and bottom centre-right zones most closely represent the original rock texture.

The sphalerite andarsenopyrite veins clearly present a problem even if they are regarded as one stage. At most points they seem to relate to the yellowish alteration, but towards the top left a sphalerite infill is in direct contact with dark wall rock and there is no yellow alteration. This alteration mismatch is again suspicion arousing. The yellow (sericite) alteration could just be matched to the arsenopyrite or more likely is related to the thin pale (pyrite) veins. Neither are particularly convincing! Several points can be observed where the dark green mineral in the host rock is partially converted to into the dull waxy yellow green sericite. Even the original rock is not what it seems as the dark green mineral represents chloritised feldspar and chloritised hornblende!

In simple language there are suspicion arousing criteria everywhere. At least three and quite probably four stages are possible. This is the wrong specimen for overprinting resolution, and the observer should stop attempting mission impossible and move to less confusing situations in the immediate vicinity. *This situation is common.* All of this fails to answer the vital question of "Which stage is gold-bearing?"

0
1 cm

◻ PLATE 19

SUSPICION AROUSING – INCONSISTENT ALTERATION – COMPLEX OVERPRINTING, PARALLEL REACTIVATION STRUCTURAL SUPERIMPOSITION – VEINS.
Sorensons tin mine, Mount Mowbray district, Queensland, Australia.

This plate has been selected to exhibit a suspicion arousing inconsistent alteration configuration.

The wall rock (bottom left) is granite, composed of feldspar (pink-orange), quartz (grey) and mica (brown). The most obvious feature is a dark vein (top) which contains very clear infill textures. These resolve themselves as dark black-brown equant crystals of cassiterite with well-developed triangular textures of infilling potassium feldspar (pink) and quartz (grey). The alteration pair that links to the infill is clearly the pink potassium feldspar, seen to alter preexisting granite feldspar and possibly quartz (top left).

A problem immediately arises as the sequence from central infill to unaltered granite is obviously different in the top left hand corner to that in the bottom right. This is a classic inconsistent alteration configuration raising the suspicion that two stages are involved. A very careful examination of a series of minute veins paralleling the main infill structure (best seen mid left) composed of quartz, feldspar and rare cassiterite reveals that they actually cut the anomalous grey alteration zone. The grey alteration zone (mid to bottom right) is in fact a previous alteration which has been overprinted by the feldspar stage. It is composed of silica (grey) and sericite (dark grey-brown iron-stained). The infill pairing to this alteration is not visible on this plate, but is confirmed as silica elsewhere on the property. The central infill zone of cassiterite-feldspar quartz actually has two different wall rocks. These are the granite as described above, and the previously altered silica-sericite rock (bottom left). The feldspar alteration within the latter has selectively picked out the sericite on the fringes.

A quartz infill vein (bottom left) without any associated alteration may represent a third stage (untimed).

0
1cm

PLATE 20

SUSPICION AROUSING – INCONSISTENT ALTERATION, COMPLEX OVERPRINTING, ALTERATION OF ALTERATION. STRUCTURAL SUPERIMPOSITION – VEINS.

Albite alteration, Mallee Gap Creek, Cloncurry region, Queensland, Australia. (Specimen provided by P. J. Williams).

This rock has been included as it exhibits three overprinting stages and gives a very good small-scale example of the very common situation where there is an asymmetry which gives the clue to separating an apparently single stage into two.

The wall rock (bottom, centre) is composed of fine-grained albite (grey) with occasional pink (feldspar?), white (quartz), and dark(?) spots. It seems to be a species of altered metasediment.

The most obvious stage is the earliest dark vein (bottom left), which is strikingly overprinted by the white/pink materials (top right). With very careful inspection the dark vein can be divided into two different domains. Centrally located, vague zones, of dark black (amphibole) infill, and surrounding zones of slightly less dark alteration (silica, amphibole?).

The overprinting materials (white/pink) clearly exist as alteration, with the whiter (albite) occurring as an alteration of the host rock. At first sight the white (albite) seems to occur as a fringe to the pink material and an initial interpretation would assess this as a zoned alteration front. However if the observer carefully tracks all around the front, there is a point (centre-top) where the pink actually comes directly into contact with host rock with no intervening white zone. The pink material is a separate stage and occurs as an alteration of the white albite, and probably the altered metasediment.

Separating the infill zones to go with the two feldspars is very difficult, and this is usually the case in feldspathic systems.

The white albite infill is just visible (bottom right) where it is distinguished by a slightly coarser grain size, and a creamy colour in relation to the adjacent alteration. Within the red material the writer cannot differentiate infill from alteration.

The pink-red material which looks like K-feldspar is in fact hematite-albite (P. J. Williams , pers. comm.) which serves to illustrate the problem of identifying pink alteration. The dark grey country rock is also albite and presumably part of the regional sodic alteration assemblage so prevalent in the Cloncurry region.

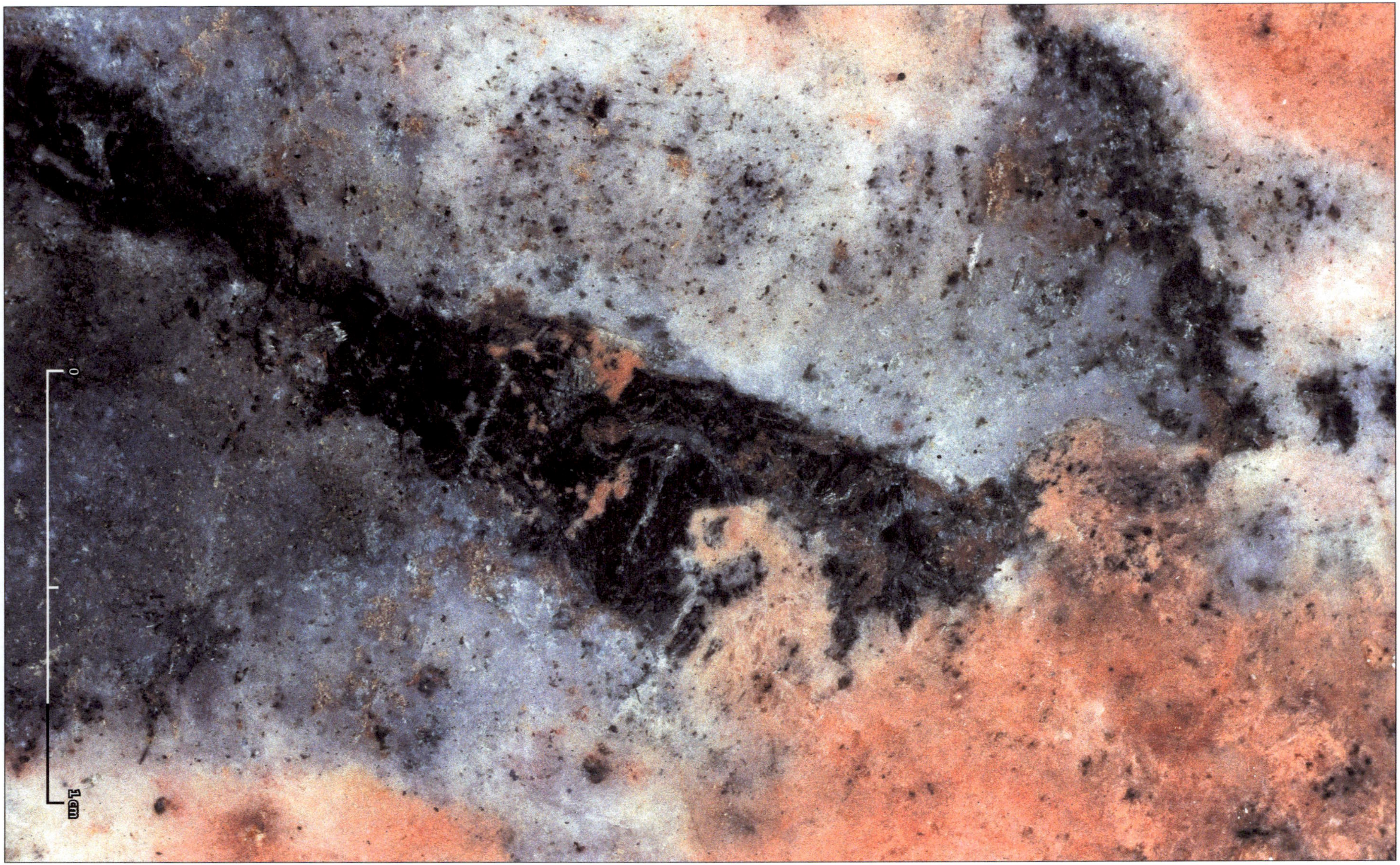
0
1 cm

◻ PLATE 21

SUSPICION AROUSING – INCONSISTENT ALTERATION, ALTERATION OF ALTERATION. STRUCTURAL SUPERIMPOSITION – VEINS. Pegmont (lead, zinc) prospect, Cloncurry region, Queensland, Australia. (Specimen provided by P. J. Williams).

This plate has been selected to illustrate subtle overprinting textures, with good examples of inconsistent alteration halos, and alteration of alteration. It will be of special interest to many readers, as it is part of the stockwork system which underlies the Pegmont lead-zinc prospect.

The host rock(s) are fine-grained metasediment(s) in the sand to silty-sand, size range. These are composed of varying combinations of minerals, pale green-grey (feldspar?) pale grey (quartz) dark black-green (chlorite, biotite?) and glassy red (garnet?).

The most obvious channelway is represented by a dark infill vein structure (centre right). The dark materials (grey quartz and black tourmaline) seem to relate to a very irregular alteration halo which is mostly white, with some irregularly distributed pale yellow components. The actual vein is impersistant, and there are blotches of dark-buff material associated with its apparent termination region in mid plate.

The left hand side exhibits two clearly different alteration styles, one of which trends vertically is mostly pale yellow with white fringes and texturally resembles the alteration described above. It seems to be controlled by minute vertical cracks, and overprints an earlier (grey, grey-white) subhorizontal alteration. This material is K-feldspar, and with close observation it appears to be replaced by the yellow-white material (sericite). This is an example of alteration of alteration, and despite the confusion relating to similar colour tones, establishes that there are two totally distinct alteration styles, and their relative timing.

It is a little difficult to relate the obvious dark infill channel (centre right) to either alteration, and the whole question of fluid channelways and ingress is not resolvable on this rock. Obviously small cracks, and variations in host rock grain size/chemistry have played a major role, with alteration predominating over infill.

This rock has been examined under the microscope and the early K-feldspar assemblages are overprinted by the sericite. Although the situation is not totally clear on the hand specimen, there are ample suspicion arousing circumstances to suggest the presence of two overprinting (and possible more?) systems, selectively utilising similar permeability zones.

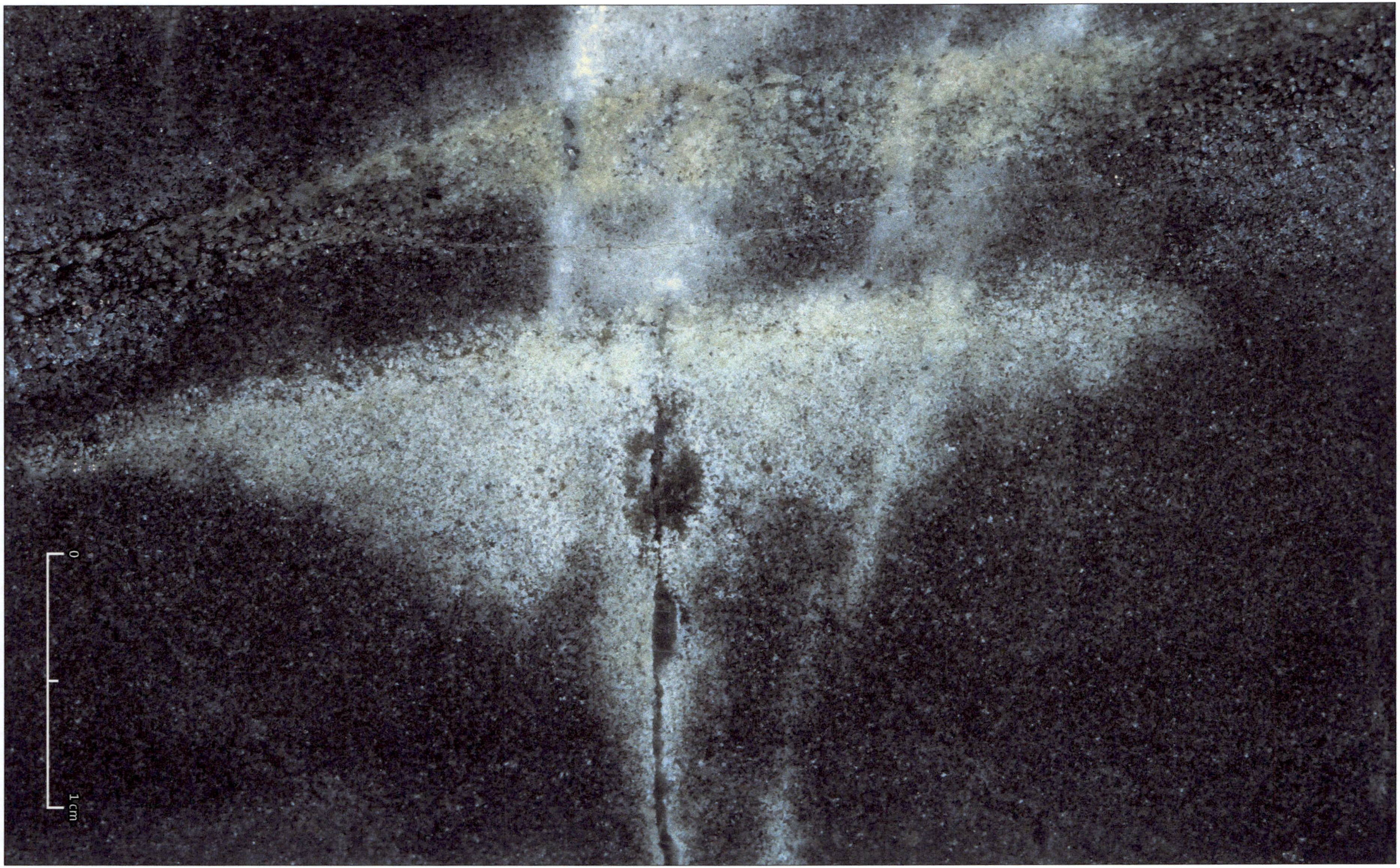
0
1 cm

◻ PLATE 22

SUSPICION AROUSING – ALTERATION OF ALTERATION, LAYER CONTROL
Tick Hill gold mine, Cloncurry region, Queensland, Australia.

This rock has been included to illustrate relatively obscure overprinting which is essentially layer controlled. It also shows a subtle example of alteration of alteration.

The specimen originates from the Tick Hill gold mine, and the pink rock is the gold-bearing unit. It is composed of albite (pale and red pink) and silica (grey). The two major components vary considerably in both format and quantity, but seem to have acquired their configuration via intense ductile deformation. The original nature of the rock is contentious, with opinions ranging from an altered representative of felsic igneous derivation though to sodium-rich lake sediments!

The situation is not made any easier by the overprinting alteration (middle to lower region). This preserves most of the original textures and results in a mixture of pale grey (quartz), pale green-white (scapolite) dark green (chlorite?) together with red brown iron-oxides. The darker zones link together and represent the main channelways controlling fluid entry. A minute pale green crack within the central portion could be an infill channel? Elsewhere on the property, the green scapolite-rich alteration can be seen taking out the entire pink rock on a major layer-controlled front.

The brown iron-oxide which occurs at many points within the plate is also an overprint, and occurs as an alteration of both the pink and green. A small cross-cutting vein is well-displayed (left hand side centre) where it crosscuts the green rock and alters the dark? and/or remnant albite (pink) component. This alteration occurs sporadically along the boundary between the two major rock units (pink and green) and from field relationships is not due to weathering.

The top of the plate just covers a change in rock type. This unit is primarily quartz (white grey) and chlorite? (dark green-black) which seems to have been affected by the iron-oxide alteration (brown). A small vein-layer of magnetite (grey-blue) is visible (top left).

0
1 cm

8 References

Dong, G., Morrison, G. W., and Jaireth, S., 1995. Quartz textures in epithermal veins, Queensland: classification, origin and implication: Economic Geology, v. 90, p. 1841–1856.

2 FROGS

This photograph is totally irrelevant, but adds a north Queensland flavour depicting two very friendly Townsvillian Bumpy Rocket Frogs. Despite close inspection normal textural terms such as overprinting, structural superimposition, or even sequential infill are considered inappropriate.

The plate comes from the surperb collection of Stephen Richards (Zoology Department of James Cook University). The correct name for these non-publicity seeking beasts is ***Citoria Inermis***.

However, the author is assured that they would prefer to be addressed as Mr and Mrs Frog.

Volume 4 – Broken Rocks · Breccia I

Volume Front Page

Complex multiple stockwork from the Grasberg (copper, gold) mine, West Papua, Indonesia.
(See also *Plate 32*, p. 215)

4.3 Systems/Rocks with Pronounced Brittle Components

1 Introduction

The subject of breccias and mineralisation is rather large and it is not possible to remotely cover the field in one slender volume. As a starting point it is perhaps best to deal with the vexing problem of breccia definition. Readers may be surprised to hear that there is no satisfactory definition of the term. This conclusion was also reached by *Laznika (1988)* in his massive, although somewhat confusing tone on the subject. Some of the problems are immediately seen in the terms sedimentary breccia and volcanic breccia. These terms are perfectly acceptable to sedimentologists and volcanologists but do not sit very well with structural/economic geologists. The latter tend to regard the term as relating to a once solid rock which is now broken and exhibits fragment rotation. To this group a sedimentary breccia is an oxymoron (see military intelligence).

Nearly all breccias involving once solid rock which has become broken with obvious fragment rotation will pass outwards into zones of less fragmentation and less rotation, and eventually terminate in a zone of fractured/shattered rock with no obvious movement of blocks (see ***Plates 3 and 11***). This tripartite pattern:

highly broken	**less broken**	**cracked**
(rotated) **(chaotic, jumbling)**	**(minor rotation)** **(mild jostling)**	**(no visible rotation)** **(simple cracking)**

is obviously a related package. However, many people do not wish to apply the term breccia to a cracked rock. Indeed, it is not unusual to see three different names given to segments of this linked tripartite zonation. i. e. intrusive breccia – mosaic breccia – crackle zone (breccia).

Given the ingrained nature of common usage, personal prejudges and different training systems, there seems to be no way out of the definition/nomenclature mad-house.

In order to maintain personal sanity the author would prefer to regard all of these phenomena as simple zones of broken rocks. However, it is inconvenient to continually call out to colleagues "Come and look at my zone of broken rock"and breccia seems destined to remain the preferred all embracing term. If we wish to be pedantic even a cracked rock will exhibit minor rotation at electron microscope scale?

The broken rocks covered in this volume are mostly those resulting from regional to local tectonism where rock units are essentially broken in situ, such that the fragments have suffered very little physical transportation. The rather loose term "*break up*" fragmentation covers the general theme. This style is undoubtedly the single most common structural control providing hydrothermal fluid conduits, and essentially involves movement along regional to local fault structures driven by large scale tectonism.

However, it also embraces the lesser recognised process of much more localised fracture development directly or indirectly relating to igneous intrusion. It is no coincidence that district scale fracture intensity increases in the vicinity of igneous intrusions, especially those related to mineralisation. Small apophyses (stocks) rising above larger scale underlying magma chambers clearly exert enormous local pressure, and a modern school of thought invokes the idea that the combination of regional tectonism and available melt, effectively raises magma pressure to easily overcome local rock stress. This pressurised melt naturally seeks out weak spots in the overlying rocks, which tend to be pre-existing faults, fractures, and fault intersections (large scale units of broken rocks). The space problem is accommodated by envisaging direct force, block uplifts, and block cut outs with huge segments effectively "sinking" in a piston-like concept (See ***Plate 33A***).

The net result is "*break-up*" fracturing focussed by local movements along pre-existing faults, and new fracturing occurring in and around the pluton. Apical zones are obviously especially prone to new fracture/fault development and it is not surprising that many stocks are surrounded by zones of increased fracture intensity, commonly at kilometre scales.

In this volume, a further form of "*break up*" is also considered which occurs when pressures in hydrothermal fluids reach sufficient magnitude to overcome local rock stress. This not only causes existing fractures/faults to dilate under fluid pressure, but also creates new cracks. The phenomenon is in wide commercial use in the oil and gas industry and is often referred to as hydrofracking. The author considers that this process is in much wider operation than might be generally conceived, and is probably much more operative in creating open space within pre-existing vein and stockwork systems than the more popular models of creating open space via rock fragmentation. Obviously both processes could be expected to commonly work in tandem.

It is worth noting in passing that the apical regions of many intrusives are also common sites for sudden/episodic generation of rapidly expanding gas streams. These gas streams are responsible for various species of breccia containing highly rounded fragments. There is also evidence of considerable fragment transportation. These are very different to the "*break-up*" breccia styles described in this text. The breccias containing rounded fragments are given many names such as hydrothermal intrusive breccia, milled breccia, pebble dyke or diatreme, and will form the subject of the next volume. However, as in all broken rock systems they also have outer zones of cracked rocks which are not only of simple break-up style, but also contribute to the high density of fracture development in these domains.

The descriptive terminology utilised for broken rocks is extremely confusing, and along with most branches of geology the structural fraternity have created their own impenetrable language which confuses and occasionally intimidates the uninitiated. In all fairness it should be noted that economic geologists have also contributed to the situation.

This is not intended to be a structural text, and the writer does indeed belong to the mildly intimidated grouping. However, some perception of terminology problems can be seen from the short list below. These terms are all applied to broken rock systems and the list is by no means exhaustive:

Tectonic breccia

Fault breccia

Fault gouge

Cataclasite (complete with protocataclasite and ultracataclasite end members)

Crush zone

Shear zone

Brittle-ductile shear zone

Shatter breccia

Crackle breccia

Mosaic breccia

Fracture zone/system

Jigsaw breccia

Stockwork

Sheeted zones

Parallel vein swarm

Cymoid loop

Fracture array

Implosion breccia

Breccia vein

Hydraulic breccia

Intrusion breccia

Tension fractures

Tension gashes

Shingle structures

Horsetail structures

Chatter links

2 Observation of Broken Rock Patterns

From the textural point of view it is recommended that all of the above terms are temporarily discarded during initial examination of hand specimens, field exposures and regional maps. Initial observation should start by simple inspection of the actual fracture/broken rock pattern in question. This will quickly reveal a much more useful perspective than trying to find an appropriate name. It is the pattern of the broken rock which is important to understanding fluid flow channelways and structural control.

It will quickly become apparent that the factors controlling fracturing/breaking are indeed complex. However several points are well worth making.

(1) Different rock types will fracture in significantly different ways.

Some rock units such as shales, schists and thinly bedded sediments already contain pre-existing parallel oriented weak points. Fractures will naturally attempt to orient or deviate to parallel the fabric (see ***Plate 19***).

Granular rocks such as garnet and garnet-pyroxene skarns tend to break up in round to ovoid units (see ***Plates 27–30***).

In other cases, the external stress field appears to play the major role, and fragmentation produces elongate to ovate-lens styles (see ***Plate 23***).

In fact, the combination of the numerous factors controlling fragmentation produces an almost unlimited variation in fragment shape. *It is always wise to look at the outer cracked zone*, as the outer fracture pattern gives a good guide as to the fragment shapes to be expected in the central dislocated rubble zones (see ***Plate 23***).

(2) The majority of fractured-broken rock zones are sites of multiple disruption, and consequently a careful examination will reveal different generations of breaking contributing to the overall pattern (see ***Plates 18, 19, 23, 26, 29, 30, 31 and 32***). This common situation simply reflects the fact that once a weak point has been established any further stress will naturally focus in and around the same place (the weak link in the chain). This is the fundamental reason why overprinting is so common, with 3–6 phases of breaking and fluid introduction following the same channelway. It is also the reason why so many vital structural controls are not recognised. The attendant infill and alteration effectively masks the multiple structures and regrettably, many structural and economic geologists do not try to unravel mineralised rocks in channelway terms. It would be fair to state that most economic geologists are not five star structural observers, and equally most structural geologists are not good observers of overprinting in altered, mineralised rocks.

(3) At slightly broader scales (air photos, satellite images) it also becomes quickly apparent that a major factor controlling breaking patterns is pre-existing structural weakness. In short, the inhomogeneity exerts a strong control creating major deviation from the standard models derived via laboratory testing of small scale homogeneous samples (blocks of marble, granite etc). Where igneous intrusions are involved, their emplacement is usually controlled by pre-existing structures. The upward ascent of cusps and stocks from the main batholith tends to accentuate the controlling fractures, particularly above the apical zones. The whole process sets up a pattern of more intense fracture around the stock which is repeatedly reactivated as a multistage hydrothermal cycle evolves.

In the same context it is always important to realise that broken rock systems will vary enormously along the relevant zone. This is again a reflection of numerous factors, with the inhomogeneity of the medium being traversed playing a major role. Brittle styles may dominate at one point (dominated by angular broken rock patterns) and abruptly change to oriented brittle-ductile shear patterns at another. In fact the combination of breaking patterns is almost unlimited, giving all the more reason to actually document the pattern rather than apply a seemingly convenient term.

(4) The problem of scale presents even further difficulties and in terms of mineralisation it usually transpires that fracture-fault-veins observed in normal scale exposures are actually part of a much larger local (or even regional) pattern. However it is a little difficult to assess this at the single exposure scale. It is usually best to adopt the approach that the visible local scale of fracturing will be a smaller scale version of the larger pattern (see ***Plates 1, 6, 7, 8, 9, 12 and 13***)

3 Approach to Understanding Broken Rock Patterns

The major problem in assessing any broken rock situation lies in adjusting thought processes to an appropriate scale. The field observer usually works at outcrop scale, and it is normal practice to race up to the interesting exposure and commence observation.

A similar process occurs with slabbed specimens where the intent is to work out what is going on with a particular rock. Whilst all of this is very necessary, it tends to distract the observer from broader scale thinking. Observers should always remember that any particular broken rock will inevitably be part of a much larger system. Thus, a crack style pattern at rock scale may relate to an adjacent fault at outcrop scale, which in turn may relate to a localised fault system at the 100 m scale. The fault system will relate to a local fault pattern and this in turn may relate to a regional pattern of major scale rock breaking. Taken to full scale the entire planet consists of broken rock units! From a practical point of view it is always wise to ask the question "how does this pattern relate to the next scale up?" This goes hand in hand with the question of "why is this rock broken?"

This text is written from a small scale rock perspective, but in order to assist the beginner, commences with a series of broader field perspectives. These illustrate a few of the more common breaking patterns and have been specially selected to give perceptions e.g. of scale. ***(Plates 1–16)***

A recommended field approach is to progress along a check-list style of observational and mental processes.

STEP 1.
Find the broken rocks.
This is best achieved by standing back:

(a) Always take a long range view before approaching the zone of interest (i. e. from the edge of the pit or around 30 m back from small exposures).
(b) Simply look at the rocks and classify into:
  1. rocks looking knobbly, cracked, broken (see ***Plate 3***)
  2. rocks looking relatively intact-unbroken (see ***Plate 3***)

The observation should also include a visual colour scan, with iron-rich/clay-rich zones receiving particular attention (see ***Plates 2–3***).

STEP 2.
Observe and contemplate the broken rocks from various scale perspectives (1 m areas, 10 m areas, etc.).
Within this text some good examples of scale perspectives are given by ***Plates 3, 4, 5, 6, 7 and 12***.

(a) pick out the major pattern (scale 1)
(b) pick out the minor linking patterns (scale 2)
(c) remember – the entire pattern is likely to belong to something bigger than the scale of outcrop, such as a wide major shear zone or a localised set of structures, and also relate to a regional pattern.

STEP 3.
Remember that the broken rock system is most likely to be a result of multiple overprinting events.
This will be especially true in mineralised examples. The best approach is to look carefully and collect large samples which can later be slabbed and inspected under laboratory style conditions. At least three generations of fracture and hydrothermal fluid introduction can be anticipated (see ***Plates 21, 26, 29, 30 and 31***).

Another major step in looking at broken rocks from a mineralisation perspective is to temporarily discard some of the simplified concepts necessarily instilled during introductory tertiary training. It takes very little experience to realise that faults drawn as single straight lines on maps are much more complex in detail and it is even more dangerous to hang onto the attractive divisions of ductile, brittle-ductile, and brittle as necessarily representing depth zones within the earth's crust. The concept obviously has some truth, but it is extremely common to find shades of all three within a single exposure. Most high strain zones in relatively solid rocks resolve the stress by shearing and/or breaking in a system of fault/crack structures.

A typical picture is a series of faults with each strand of the system moving at slightly different rates (high strain zones) such that the portions in between are forced to break in a secondary pattern to accommodate the strain. The net result is a combination of shear and tensional breaking which occurs at a range of scales.

4 Arrangement of Plates – Design Philosophy

Philosophy

Considerable difficulty has been encountered in arranging the illustrations in any logical format due to the problem of categorising broken rocks. In mineralised domains they have inevitably been affected by 3–5 different tectonic events, which may be expressed in entirely different ways, and formed under totally different regimes.

For instance, it is almost the norm for an early formed ductile or brittle-ductile shear zone to be reactivated at some later time by brittle dominated events associated with hydrothermal activity. Possibly dilation and further crack extension via pressurised fluid further adds to the picture!

Similarly, it is again almost the norm for deep-seated shear-related structures occupied by barren quartz of the slate belt/greenstone gold style to be overprinted by brittle crack styles at much later times, with the gold relating to these later events (see ***Plate 33***).

After considerable thought, the pattern given below has been adopted. Hopefully the critical reviewers will forgive the inevitable overlaps. Considerable effort has been devoted to plate descriptions and cross-referencing. The whole purpose of the series is to encourage readers to look at rocks from a field and/or hand lens perspective. It is the texture that is all important and frequently overlooked by the tendency to jump from outcrop to thin section scale without slabbing.

General Pattern of Presentation

4.1 Problems of perception and general patterns at the larger scale (outcrops and small scale slabs representing miniatures of large scale rock fracture/breaking patterns).
(***Plates 1 - 20***)

4.2 Dominantly brittle-ductile and ductile shear styles. Overprinted by later hydrothermal events.
(***Plates 20 - 23***)

4.3 Dominantly brittle styles, and complex hydrothermal overprinting with emphasis on different breaking patterns and possible fluid pressure effects.
(***Plates 24 - 33***)

4.4 A miscellaneous section!
(***Plates 34 - 36***)

4.1 General Patterns of Broken Rocks and Problems Involved with Recognition, Scale Perceptions and Overprinting

◻ PLATE I

Patterns (shear + brittle adjustment, concentrated fracture)
Problems (scale perceptions)
Antisocialist tin mine, Herberton, Queensland, Australia.

This plate contains many instructive features.

The rock is a broken, silicified metasediment which has been refractured. The silicification has created a reasonably uniform competent rock. The new fracturing is outlined by rusty iron-oxides resulting from percolating oxidised meteoric water (weathering). A little clay (white-infill) of uncertain timing and origin is also present. The clay occupies cavities which precede the larger open cavities. The big open cavites (dark) represent zones where fragments have fallen out during slabbing, and some genuine open cracks, unfilled by iron-oxides/clays.

The seemingly random stockwork pattern is actually highly ordered. If larger fractures are separated from smaller ones a semi-vertical (NNE) branching system emerges which is linked by a complex mesh of smaller fractures. The latter are also far more ordered than suggested from initial impressions (NNE-NE, NNW-NW).

Inspection of the larger scale fracturing to the top right reveals a small scale brittle-ductile shear fabric, with elongated aligned lensoid components. It is probable that most of the semi-vertical fractures contain small elements of displacement with a major shear operating at top right.

This gives some clues as to how this very common style of pattern develops. Movement occurs as a linked series of semi-parallel fractures (shear zones), and the strain in between is accommodated by linking similarly oriented (branch) fractures which also become slip planes. However, movement along them is at different rates, as they differentially accommodate the strain. This creates a situation where the intervening rock slabs simply crack, to create the second order brittle shatter patterns with little or no movement. Hence the whole pattern is formed by varying combinations of simultaneous movement and cracking. (Other variations on this theme are shown in *Plates 2, 3, 4, 5, 6, 7 and 18*).

Zones of high fracture density are apparent, and potential fluid flow patterns are easily visualised.

The meteoric fluids which created the iron-oxides, operated at low pressure and simply filtered through. As a mental exercise it is not difficult to imagine the effect of a high pressure fluid. This would accentuate the fracture pattern, dilate selected channels, create new fractures, and could ultimately create a more visible stockwork system (see *Plates 18, 24, 31 and 32*). On the other hand, any major wall rock alteration would serve to conceal the fracture pattern.

A further valuable exercise is to visualise the plate at a regional scale perspective – say 20 × 35 km. The situation could easily match a portion of the Andean zone of South America, with major N-S regional faults and their interconnected secondary fracture zones. The Andean zone is underlain and intruded by major igneous batholiths, and major ore deposition is commonly associated with apical apophyses of magma rising above the main chambers. If a major magma chamber (20 × 35 km) were to underlie the fractured ground depicted in the plate, it can easily be envisaged where forceful stock style emplacement would be favoured. They would presumably pick out and push up through the zones of most broken ground. The sector towards the base (marked by interconnecting cavities) would be a good example. Associated refracturing controlled by pre-existing structure could easily provide channelways for associated hydrothermal fluids to form a localised vein field!

The plate could also be viewed at the 10 m or even 100 km scale, and is an excellent introduction into the concept of treating broken rocks with scale in mind. Almost any broken rock will prove to be part of a larger system which may not be obvious to the scale of observation (see *Plates 8-9*).

PLATE 1 **Patterns (shear + brittle adjustment, concentrated fracture)**
Problems (scale perceptions)
Antisocialist tin mine, Herberton, Queensland, Australia.

Plates 2–3

Patterns (shear and brittle adjustment)
Problems (visual recognition – knobbly texture, scale perception, oxidation concealment, paragenetic concealment)
Telegraph tin mine, Herberton, Queensland, Australia.

The plates opposite have been selected as typical weathered oxidised exposures of a brittle-ductile faulted broken rock system. A strongly recommended procedure is to commence the observational process with a long distance overview. The reasons for long distance perspective are:

1. The suspected broken rock zone(s) will inevitably be part of a much larger system, and it is wise to get a broad scale overview. This means observers should resist the natural temptation to rush up to areas of interest. A quiet reflective walk around the pit edge is highly recommended.
2. Broken rock zones reveal themselves via several parameters, and as mentioned previously many observers simply fail to see them. The most useful techniques are:

(a) Stand back and systematically ask the question "which areas of this exposure look broken or cracked?" More importantly "which bits look more shattered than the others (relatively disruption)?" For instance within *Plate 2*, the area to the left looks very broken (knobbly), the central zone is cracked but relatively intact, while the right hand side is extremely broken with an obvious rock fabric.

The same exercise can be repeated utilising *Plate 3* at a different scale. It is worth noting that weathered outcrops assist considerably in bringing out the knobbly/broken patterns.

(b) Combine the search for broken rocks with any obvious geological/alteration features. In general terms noticeably white and/or red zones deserve special attention. Red zones usually reflect a permeable zone, which originally contained iron-rich minerals (sulphide, chlorite, etc.) now converted to limonite via the weathering process. White zones normally reflect strong kaolinite alteration of wall rocks and/or broken rock zones, relating to acid-rich surficial water derived from sulphide oxidation.

Plates 2–3 exhibit red zones corresponding with the broken-knobbly criterion and *Plate 3* has narrow linear white clay zones. The close up photograph exhibits a structural pattern which defines an anastomosing complex of stronger fractures (high strain shear zones) linked by cracked rocks. (See also the *line drawing*).

It is extremely common for the most visible fracture pattern to represent late-space to post-mineralisation movement, superimposed over a zone of repeated tectonism. In this instance the white fractures are very late stage features (acid fluid-clay alteration/precipitation) which overprint a more strongly developed shear system (red-linear), which in turn overprints a linear milled breccia (red fine-grained rubble). Cassiterite occurs in the milled breccia. Sense of movement is difficult to define (given at least three disruptions). The overall pattern suggests a strong dextral influence at some point.

Representation of broken rock pattern (see Plate 3)

Thick lines – major fracture.

Thin lines – minor fracture.

Brown shading – highly broken/ oxidised.

PLATE 2 Patterns (shear + brittle adjustment) Problems (visual recognition – knobbly texture, scale perception, oxidation concealment, paragenetic concealment) Telegraph tin mine, Herberton, Queensland, Australia.

PLATE 3 Patterns (shear + brittle adjustment) Problems (visual recognition – knobbly texture, scale perception, oxidation concealment, paragenetic concealment) Telegraph tin mine, Herberton, Queensland, Australia. (photographed to the right of Plate 2)

◻ Plate 4

Patterns (blocky, broken rock)
Problems (visual recognition, gossan, leached outcrop, infill recognition, paragenetic concealment)
Isobella (siver, lead, zinc) mine, Herberton, Australia.

This plate may dispel some of the preconceived visions concerning broken rocks, especially with respect to fault zones. The plates opposite are well worth inspection as the brecciation and consequently the mineralisation style was expensively misidentified. The brecciation was not even recognised!

In terms of the "rules", a simple stand back glance at *Plate 4* should reveal a very "knobbly" exposure with obvious iron-oxide concentrations signalling the probable presence of broken rocks.

The zone extending back to the vehicle is the trend of a major fault zone, which at this point traverses quartzite/meta-sandstone. The fragmentation occurred across the exposure and the blocks are relatively large. Note there is only the vaguest hint of fragment orientation parallel to the hill slope.

◻ Plate 5

Patterns (blocky, broken rock, irregular limonite gossan)
Problems (leached outcrop – infill recognition)
Isobella (silver, lead, zinc) mine, Herberton, Queensland, Australia.

The lower close up reveals a similar picture to the stand back view above, in that fragments are not immediately obvious. The key factor here concerns the rusty zones, which on close inspection are composed of limonite with numerous large quartz crystals. The latter provide the valuable information that the iron-oxide rich zones are in fact large holes once infilled by quartz and an iron-rich sulphide (now oxidised – gossan).

The small photograph below is a piece of primary ore, which is composed of infilling iron-rich sphalerite (black) with quartz crystals passing into silicified wall rock and a quartz-rich host meta sediment.

The underground exposures reveal infilled cavities at the 0.5–1 metre scale. It is interpreted that the very brittle rock has fractured/faulted into large blocks with either block rotation and/or high pressure fluid introduction creating large, irregularly distributed cavities. Sphalerite and galena are the main infill minerals. The large cavities gave a false impression of massive sulphide during drilling and despite the infill character they were judged to be exhalative. Unfortunately, the isolated cavities were erroneously linked between drill hoes for ore reserve calculations.

Primary ore relating to Plate 5. Quartz (white) sphalerite (dark) wall rock (grey).

PLATE 4 Patterns (blocky, broken rock)
Problems (visual recognition, gossan, leached outcrop, infill recognition, paragenetic concealment)
Isobella (silver, lead, zinc) mine, Herberton, Australia.

Patterns (blocky, broken rock, irregular limonite gossan)
Problems (leached outcrop – infill recognition)
PLATE 5 Isobella (silver, lead, zinc) mine, Herberton, Queensland, Australia.

◘ Plates 6–7

Patterns (shear + brittle adjustment)
Problems (visual recognition, scale perceptions, oxidation gossan concealment, paragenetic concealment)
U. N. A. Balgammon (copper, tin) mine, Watsonville, Queensland, Australia.

The plates opposite, and the line drawing below have been selected to again illustrate some of the problems involved in identifying broken rocks within weathered profiles, and also to impress upon inexperienced readers that fault zones are commonly quite complex in nature. The plate also serves to provide a good impression of a low angle brittle-ductile mineralised fault zone/system.

The upper plate depicts an exposure which was not correctly observed in initial investigations, and a failure to recognise the nature of the deposit proved to be an expensive error.

The presence of knobbly rocks, partially linked to limonitic zones provides two vital observations. The process of separating very broken rocks from less broken rocks will quickly reveal a pattern similar to that simplistically outlined in the line drawing. (see *Plate 7* opposite).

This reveals a prominent set of structures associated with broken and cracked rocks which form part of a larger anastomosing brittle/shear system.

The iron-oxides (red) are accompanied by minor occurrences of copper secondary minerals (malachite, azurite) and also secondary arsenical products (scorodite). These are not visible on the plate but reflect the fact that the primary ore is composed dominantly of pyrrhotite, chalcopyrite and arsenopyrite. Minor sphalerite, galena, stannite and cassiterite are also present and the paragenesis is complex – multistaged.

The sulphide oxidation has etched out the ore structure, with the smaller rock fragments representing silicified comminuted wall rocks which were originally surrounded by infill (and alteration) sulphide assemblages. This assemblage overprints an original shear system. It is possible that later reactivation of the fundamental underlying fault system is also represented within the exposure. The system portrayed in the plates is part of a major low angle fault system, running semi-parallel to bedding and porphyry sills. (see *Plates 8–9* for larger scale equivalent).

Representation of broken rock pattern (see plate 7)
Thick lines – major fractures. Thin lines – minor fractures.
Brown shading – very broken/oxidised.

PLATE 6 Patterns (shear + brittle adjustment)
Problems (visual recognition, scale perceptions, oxidation – gossan concealment, paragenetic concealment)
U.N.A.-Balgammon (copper, tin) mine, Watsonville, Queensland, Australia.

PLATE 7 Patterns (shear + brittle adjustment)
Problems (visual recognition, scale perceptions, oxidation concealment, paragenetic concealment)
U.N.A.-Balgammon (copper, tin) mine, Watsonville, Queensland, Australia.

Representation of major elements of broken rock pattern (see plate 8 opposite)
White lines – main fractures.

◘ Plates 8–9

Patterns (complex low angle shear system, repetitive shear and brittle adjustment)
Problems (scale perception)
Nolan´s gold mine, Ravenswood, Queensland, Australia.

The two plates opposite depict a good example of a low angle brittle-ductile, shear/fracture system, which certainly makes the point that an individual fault/fracture system ultimately has to be viewed in the context that it will be part of a much larger system. Fairly obviously, it is not possible to stand back far enough to give a full overview despite the extensive excavation.

The bulk of the fault/fracture system has been well outlined by oxidised meteoric water precipitating "limonite" along the main fracture surfaces. The spaced low angle pattern with fragments (cracked rocks) in between, is partially outlined in the line drawing above, and seems to reflect a reactivation of an almost identical underlying pattern containing complex sulphide-dominated gold-bearing veins.

The low angle pattern seems to link to a major vertical structure, some elements of which are just visible to the far left.

The close up photograph (***Plate 9***) is in essence a small scale example of the bigger picture, with spaced, mildly anastomosing faults/fractures angling across the field of view. Some shear style structure is visible (top left) and the cracking of rock at right angles to these is clearly visible. This is interpreted as tensional style stress release within intervening blocks caught between variable movements on the low angle structures.

Note how the recent oxidation (reddish iron-oxide) has overprinted pre-existing structures.

The Nolan´s gold occurrence is some 11 million tonnes at 1.7 g/t Au, associated with base metals. A long 8–12 stage paragenesis is involved, with repetitive rock fracturing/faulting. This is partially concealed in the oxidised domains.

PLATE 8 Patterns (complex low angle shear system, repetitive shear and brittle adjustment)
Problems (scale perceptions) see opposite line drawing.
Nolan´s gold mine, Ravenswood, Queensland, Australia.

Patterns (complex low angle shear zone, repetitive shear and brittle readjustment – see also plates 7 and 8)
Problems (scale perception)
PLATE 9 Nolan´s gold mine, Ravenswood, Queensland, Australia.

Representation of broken rock pattern in Plate 10 – opposite.
Sense of shear suggested by major elements (top right) is dextral.
White lines – main fractures.
Brown shading – very broken/oxidised zones.

◘ Plate 10

Patterns (brittle-broken faults)
Problems (repetitive fracturing mineralisation)
Grasberg (copper, gold) mine, Heavy Sulphide Zone, Papua, Indonesia.

The plate and line drawing illustrate a typical unmineralised fault pattern in predominantely brittle mode. Viewers are invited to inspect for open space (very limited) and to visualise just how a hydrothermal fluid would permeate through this broken style. Even a fluid under high pressure is unlikely to form a vein. Small scale patterns of highly broken rocks to less broken are visible throughout. The major linear zones (thick lines) probably represent a late reactivation overprint cutting through the rubble. The fragments contain several stages of mineralisation/alteration formed prior to any of the obvious disruption. Note the fragment shapes.

◘ Plate 11

Patterns (central rubble, marginal jostle, outer fracture)
Problems (repetitive fracturing obscures paragenesis)
Grasberg (copper, gold) mine. Heavy Sulphide Zone, Papua, Indonesia.

An example of an unmineralised fault zone showing brittle fracture intensity and fragmentation increasing towards the central region. Note larger extent of peripheral fracture. The disruption postdates several stages of mineralisation/alteration (See also *Plate 10*).

PLATE 10 Patterns (brittle-broken faults) and problems (repetitive fracturing – mineralisation) Grasberg (copper, gold) mine, Heavy Sulphide Zone, Papua, Indonesia.

PLATE 11 Patterns (central rubble, marginal jostle, outer fracture) and problems (repetitive fracturing obscures paragenesis) Grasberg (copper, gold) mine, Heavy Sulphide Zone, Papua, Indonesia.

Representation of main elements of broken rock pattern in Plate 12.

Thick lines – major fractures.

Thin lines – minor fractures.

Plate 12

Patterns (shear and brittle readjustment)
Problems (visual recognition at different scales, overprinting of previous broken ground, mineral paragenisis concealment)
Boska tin mine, Herberton, Queensland, Australia.

The top plate (opposite) has been selected to again illustrate that many faults are not single isolated zones, but are commonly composed of linked networks of individual components. The major components of this brittle style system all show minor amounts of crushed rocks at a close-up scale, and probably mark sites of minor movement. The important point is that the intervening zones have been placed under stress, and have responded by simple but extensive cracking. The crack (broken rock) pattern for a portion of the system is illustrated above, where it can be observed that more extensive cracks are interlinked by smaller cracks (a cobweb effect). This is a very typical situation for large scale fractured ground preparation. It is not difficult to imagine the effect of introducing either a low or high pressure hydrothermal fluid with major alteration plus or minus additional hydrofracture – and potential stockwork formation. If alteration prevails it could become very difficult to discern the original structural controls. In fact, the unmineralised fracture pattern in the plate overprints a completely altered rock (sericite-silica) with minor isolated infill zones (quartz-cassiterite) suggesting this is not the first time that this zone has fractured in this manner! (see also *Plates 10, 11, 12, 25, 26, 29 and 30*).

Plate 13

Patterns (typical brittle breaking)
Problems (overprinting previous broken ground, mineral paragenesis concealment)
Puquqaqa copper prospect, Peru.

This plate (opposite) has been paired with the one above to illustrate a very similar style of unmineralised brittle fault-fracture (broken rock pattern) again superimposed on previously fractured-altered (argillised) igneous rocks. The new pattern is partially outlined by limonite alteration via surface oxidation (oxygenated meteoric water). The argillisation is a fringe zone effect relating to a very large (kilometre) scale "intrusive-diatreme" copper mineralised system. The ever diminishing, crude cobweb pattern can be seen as the rocks have cracked to take up strain between the differential movements of the main movement zones (thicker lines). Considerable amounts of small scale rubble with fine clay gouge are present. It is possible (probable) that two separate unmineralised systems are represented, with the cracks now outlined by limonite belonging to a more recent small scale reactivation. This style of brittle ground preparation is similar to the situation discussed above, and illustrates small scale detail.

PLATE 12 Patterns (larger scale broken-shear + brittle fracture readjustment)
Problems (visual recognition at different scales, overprinting of previous broken ground, mineral paragenisis concealment)
Boska tin mine, Herberton, Queensland, Australia.

PLATE 13 Patterns (typical brittle breaking)
Problems (overprinting previous broken ground, mineral paragenesis concealment)
Puquqaqa copper prospect, Peru.

Plate 14

Patterns (sheeted vein, dilated)
Problems (dilation)
Mt Carbine tungsten mine, Queensland, Australia.

A spectacular example of parallel sheeted fracturing now occupied by major quartz veins. The latter contain substantial wolframite and were mined in recent times via an open pit selective mining process. The quartz-wolframite components were separated via automated colour sorting subsequent to crushing. The wall rocks are fine-grained siliceous metasediments. This style of subparallel sheeted veining is extremely common in quartz wolframite concentrations throughout the world. Numerous examples form directly above underlying granite cupolas. The lower and upper parts of the system (1–2 km scale) feather out, with maximum vein widths occurring in the central and lower central sectors. The veins are clearly infill dominated, and high fluid pressures are suspected to play a major dilational role. The fracture style seems connected directly to the probable forceful intrusion of the underlying igneous apophysis/stock. Movement along fractures is minimal and it seems as if an array of brittle parallel cracks formed in relation to the local stress domain, usually within rocks that are relatively isotropic. An analogy of driving a blunt chisel into an unfractured rock might be appropriate?

Plate 15

Patterns (sheeted veins, dilated)
Problems (dilation)
Sahakit tin mine, Phuket, Thailand.

A similar example to ***Plate 14*** of parallel sheeted style fracturing, but with quartz-cassiterite veining. The host rock is argillised fine-grained granite. The rock face is sluiced, and cassiterite is ultimately extracted via sluice boxes/shaking tables (see also ***Plate 24F***).

Plate 16

Patterns (stockwork)
Problems (scale perception, origin)
Stackpool tin mine, Ardlethan, New South Wales, Australia.

The intensely brittle fractured zone now accentuated by surficial weathering and mining activity would certainly qualify for the term stockwork. The large zone of shattered rock forms the marginal zone to an intrusive style hydrothermal breccia pipe. These are thought to form via the escape of rapidly expanding gas. Not surprisingly, many of these columns of milled fragmented rocks are surrounded by zones of intense brittle fracture. The latter form perfect sites for permeation and mineral deposition via later generations of more conventional hydrothermal fluids.

PLATE 14 Patterns (sheeted vein, dilated)
Problems (dilation)
Mt Carbine tungsten mine, Queensland, Australia.

PLATE 15 Patterns (sheeted veins, dilated)
Problems (dilation)
Sahakit tin mine, Phuket, Thailand.

PLATE 16 Patterns (stockwork)
Problems (scale perception, origin)
Stackpool tin mine, Ardlethan, New South Wales, Australia.

◻ Plate 17 A B

Patterns (fractures/faults in relation to folds)
Problems (scale perspectives, different fluid generations)
Unlocated specimen, Hodgkinson goldfield, Queensland, Australia.

The specimen has been selected to give a visual understanding of rock breaking associated with fold development. The pre-folding package consisted of fine-grained, siliceous sediments (mudstones) which are now represented as siliceous metasediments (grey-cream) and quartz veins (pale to dark blue grey).

The rocks probably contained a pre-existing cleavage (not visible) and the main cleavage structure now visible is a spaced differentiated crenulation cleavage (the dark cracks running approximately parallel to the axial plane of the fold with mildly radial developments towards the basal sector). In detail these are actually subparallel sets of brittle cracks which are now filled with grey quartz, suggesting that they have been utilised by the introduction of pressurised hydrothermal fluid.

The larger pale to dark blue-grey quartz veins have clearly been folded with mild crenulation visible in some small scale examples (lower *Plate 17B*). The competent quartz has responded via brittle fracture and many elements of this can be seen from the faint darker structures running at right angles to the vein and radially around the fold axis. This weakness has been exploited at some point by a stage precipitating white infill quartz (centre right). This particular hydrothermal event can be seen at many points within the rock (top left, centre right) and obviously occurred prior to the shear style disruption (dark layers).

The flanks of the fold ultimately become areas of extremely high strain, particularly with reference to lower strain domains towards the centre and apex. The net result is that the fold breaks and shears develop along the flanks. These are well displayed on the plate as antasomosing dark wavy zones (top right) which displace segments of pre-existing quartz veins (both early blue and later white) as well as shearing out, stretching, crushing, and grinding rock fragments. Under a 15× hand lens, the rock fragments within the shear zones are altered (buff to green) and the entire shear network is dark green, suggesting the passage of a hydrothermal fluid (low pressure) causing sericite/chlorite? alteration.

It can be seen that folding introduces a combination of brittle and ductile rock breaking factors simultaneously. Quite clearly any further tectonic disturbance (such as granite intrusion) will reactivate any or all of these with focussed rebreaking. Development of regional scale faults (either brittle or ductile) will tend to have major segments following the fold limbs (that is the country rocks will break along pre-existing weaknesses). Similarly the introduction of a high pressure fluid at any time will create dilation and further associated rock fracture. Lower pressure fluids will permeate along both brittle and ductile structures. Obviously, the combination of regional/local major fault movement of structures already containing hydrothermal fluids will create massive overpressuring/dilation/rock breaking. Many veins, vein systems, are controlled in this manner, and there are many lessons to be learned from this plate. Again it is wise to think in different scales, and make the plate 0.5 km across, changing the folded quartz vein to a quartzite (see also *Plates 18 A, B, 19 and 33*).

PLATE 17 B **Higher magnification insert from Plate 17A Hodgkinson goldfield, Queensland, Australia. (Specimen provided by G. W. Clarke).**

PLATE 17 A Patterns (faults/fractures in relation to folds) Problems (scale perspectives, and different fluid generations).

◻ Plate 18 A B

Patterns (brittle ductile combination)
Problems (timing relations)
Mt Isa (lead, zinc, copper) mine, Mt Isa, Queensland, Australia.

This specimen has been included to illustrate two different styles of rock breaking, which clearly occur at different times. It is also a common pattern.

The specimen is composed of a fine-grained, hard dark siliceous unit (central zone) flanked by highly strained soft, fine-grained paler (micaceous) schistose materials. The dark unit is actually a tuff marker horizon (silica ± K feldspar), and the paler altered unit was originally a dark grey calcareous shale (Urqhart Shale). The latter may contain elements of an early ductile shear fabric.

The competent siliceous layer has fractured in a brittle manner to give an irregular high angle tensional style response. This is fairly typical of competent units caught up in high strain situations (i. e. minor sandstone units in shales). The softer (aluminosilicate) rocks respond via ductile shear with the competent zones simply shattering (strain partitioning).

Early stage hydrothermal fluids have permeated into the cracked zone both via the broken contact with the less competent rock, and through the cracked zone. The white-grey vein materials are dominantly quartz ± dolomite? and within the close up photograph, signs of layering and coarsely granular crystalline materials indicate infill. It seems probable that the fluid caused the original crack system to dilate. There is no associated siliceous alteration. A paler grey zone beneath the obvious central suphide veins (*close-up plate*) looks like some form of prior alteration of the darker unit and actually effervesces vigorously in dilute hydrochloric acid – carbonate alteration?

The pyrite rich (pale yellow) materials clearly follow a totally different arrangement, and occupy minor cracks, small scale faults, and schistosity domains most of which are essentially parallel to the ductile/shear fabric. Major disruption has occurred along and/or close to the incompetent/competent rock junction. It represents the "weak link" in the chain and ultimately ruptures. It is a little difficult to distinguish infill from alteration in the pyrite domains, although clearly sulphide alteration has occurred. The tiny irregularly spotted sections in this schist denote sulphide-spot alteration as fluids permeated along the fabric.

Minor late stage brittle cracking overprints the sulphide stage and contains a minute white late silica ± carbonate? infill. The redder stains towards the peripheries, are oxidation/iron-oxides formed from percolating ground water.

This type of situation is very common at a variety of scales. For instance, igneous sills and dykes in layered metasediments are common sites for focused fracturing and shearing in and around their competent/incompetent contacts (see ***Plate 7***). At a broader scale, the situation within the photographs above might relate to fold limb strain focusing analogous to that shown in ***Plate 17A, 17B***?

PLATE 18 A Patterns (brittle-ductile combinations involving a competent layer set in less competent surroundings)
Problems (timing relationships)
Mt Isa (lead, zinc, copper) mine, Queensland, Australia. (Specimen provided by W. P. Laing).

Higher magnification picture of a portion of Plate 18A to show timing relationships.
NOTE: This pattern is very common, commonly involving porphyry/basaltic sills and dykes within less competent, layered units. (e.g. the Mt Charlotte gold stockwork, Kalgoorlie, Western Australia).

PLATE !8 B Mt Isa (lead, zinc, copper) mine, Queensland, Australia. (see also Plates 17A, 17B and 19)

◘ Plate 19

Patterns (layer parallel shearing at small scales, brittle-ductile combinations)
Problems (visual recognition, paragenetic timing of different events, interpretation)
Mt Isa (lead, zinc, copper) mine, Mt Isa, Queensland, Australia.

This rock has been included to illustrate a combination of three problems which commonly confront any observer of "mineralisation textures" from a broken rock perspective.

PROBLEM 1 – Actual recognition/observation of the textures. In this case the original texture in the underground rock face reveals almost nothing. The slabbed rock brings out a few textures, but the full visual picture has only been presented after careful treatment with different size grinding powders.

PROBLEM 2 – Recognition of multiple overprinting hydrothermal stages. A good understanding of infill, alteration and overprinting textures is required to interpret structural controls and paragenesis.

PROBLEM 3 – Given the above, an interpretive stage still leaves room for errors/ambiguity and it may be necessary to follow-up with a complete suite of rocks and targeted microscopy. The above problems are always compounded when permeability channelways are liable to be layer parallel, and there are substantial post-mineralisation deformation effects! All of the above apply to the major Mt Isa lead, zinc, copper systems and their related cousins throughout the northern region of the Mt Isa province (Century, McArthur River, Hilton, George Fischer etc).

The rock depicted is from one of the Mt Isa lead-zinc orebodies (see also ***Plate 18A, 18B***). The pale green layer (left) is a "tuff marker bed" composed of fine-grained silica, feldspar ± carbonate, and the remaining metasediments are fundamentally variations on carbonate-rich siltstones and shales.

In terms of mineral identification:

1. The prominent white veins in the tuff marker bed and elsewhere are essentially coarse-grained dolomite ± quartz.
2. The prominent bright spots (left) are pyrite (yellow).
3. The pale brown material dominating the right hand side is sphalerite.
4. The intermediate bright buff coloured material which is prominent in a centre left band is pyrrhotite.

In terms of broken rocks and paragenesis the white dolomitic veins are the earliest feature. They are clearly infill in character with no obvious associated alteration. Those situated in the tuff marker bed occupy dilated cracks (fluid pressure dilation?) and a similar origin can be postulated for the segments of vein materials running semi-parallel or at low angles to layering situated within the pale brown sphalerite band(s). An impression is rapidly gained that the latter have been disrupted – faulted.

Examination of the sphalerite reveals two major textural domains:

(a) Bedding parallel and/or low angle to bedding zones of fine-grained massive sphalerite.
(b) Lensoid zones in between the above containing elongate blebs-spots-discontinuous tiny veins.

There are numerous examples of a combination of the above which is interpreted as small scale discontinuous layer replacive sulphide alteration (see ***Plate 36***).

Taken as a total package, the entire picture can be interpreted as an anastomosing micro shear system which has formed a channelway for hydrothermal fluid flow (see ***Plates 20 and 21***). Sphalerite formation is almost totally in replacive mode which blurs the original shear structure. The total to partial alteration reflects the permeability of the original channelways, and given that the unsheared rock was probably originally very carbonate rich, a classic carbonate replacement model seems probable. The pre-existing coarser-grained vein dolomite is less prone to alteration, but on close inspection numerous sphalerite alteration spots occur within the veins. The shear style hypothesis fits very well with the pre-existing vein dismemberment.

The pyrrhotite has photographed very well and can easily be distinguished (lighter yellow colour) from the sphalerite (pale brown). A very similar picture emerges (solid and spotty zones). This is well seen at far left where minute bedding parallel to low angle massive zones anastomose along the layer with discontinuous spot-lens style sulphide alteration occurring in the intervening rock-lensoid zones. A broken dark rock layer reinforces the picture of small scale movement. The pyrrhotite to the centre and left occurs in a wide variety of layer parallel and crosscutting flecks giving a strong impression of overprinting the sphalerite and again picking out the original shear fabric.

The pyrite distribution is more difficult to interpret. The isolated spots (left) exhibit good crystal shapes but there is no obvious channelway relationship. Their timing is thus obscure, possibly early? (diagenetic?).

PLATE 19 **Patterns (layer parallel shearing at small scales, brittle-ductile combinations) Problems (visual recognition, paragenetic timing/overprinting interpretation). Mt Isa (lead, zinc, copper) mine, Mt Isa, Queensland, Australia. (Specimen provided by G. W. Clarke). (See also Plates 17 A, B, 18 A, B, 20, 21 and 22).**
At a bigger scale the situation could be analogous to Plate 17 A, B with a fold limb involvement.

4.2 Systems/Rocks with Pronounced Shear (Brittle-Ductile to Ductile) Components

◻ Plate 20

Pronounced involvement of shear components (overprinted by successive hydrothermal events)
Starra (copper, gold) mine, Selwyn, Queensland, Australia.

This plate has been selected to show a reactivation "breaking" pattern controlled by a pre-existing ductile shear fabric. This is also a very common situation where mineralisation overprints pre-existing shear style fault fabrics.

The photograph is over-exposed to bring out the texture. The pale grey to dark grey rock fabric is black in reality and the elongate-lensoid nature of ductile "flattened-stretched" fragments is pronounced, creating an anastomosing effect. These are now composed of chlorite ± quartz. It is quite possible that an early biotite alteration may have been present given the nature of the system? (occasional dark greenish tinge).

The early shear fabric is now overprinted by an another anastomosing pattern outlined by grains, granular aggregates, and minute veins of pale silver hematite. These are not deformed and outline the pathway which controlled hydrothermal fluid entry. The hematite is actually an alteration of pre-existing magnetite which is recognisable from the crystal/grain shapes. Fairly obviously, this is a re-utilisation of the existing shear fabric. It is not clear whether or not there is re-shearing or just re-cracking involved both are suspected?

The granular magnetite/hematite is formed mostly as an alteration component, which tends to obscure the original channelway structures. Some of the minute cracks are probably infill dominated, although it is fairly clear that open space is minimal.

(See also *Plates 2, 3, 6, 7, 8, 9, 17, 18, 19, 21, 22, 34 and 35*).

◻ Plate 21

Pronounced involvement of shear components (overprinted by successive hydrothermal events)
Starra (copper, gold) mine, Selwyn, Queensland, Australia.

This specimen is an excellent visual example of multiple usage of an original brittle-ductile shear fabric, and exhibits a very typical reactivation, overprinting fracture system.

The darker pink fine-grained siliceous ± hematite host rock has been part of a major ductile shear fabric. The initial fabric has controlled the entry of high temperature fluids which resulted in significant albite alteration (paler pink). There is no discernable infill and hence a relatively low pressure fluid involvement is inferred.

At some subsequent time the inherited shear/albite fabric has been fractured in a brittle fashion, presumably as a result of stress refocusing along the original fault zone. The net result is a subparallel lensoid-ovoid anastomosing style of fragmentation. In the plate opposite, this is a species of crackle zone with very little evidence of rock breaking (gouge, rock fragments) or new shear fabric.

An introduction of hydrothermal fluid presumably with a reasonably high pressure has forced apart the crack system and eventually precipitated fine-grained calcite (grey), magnetite (black) rare quartz and possibly biotite (chlorite altered)? This stage has no obvious accompanying alteration. Although not totally clear the possibility of an even later carbonate infill stage is suggested by the whiter central vein.

Brittle style fracture superimposed upon more ductile styles is a very common situation.

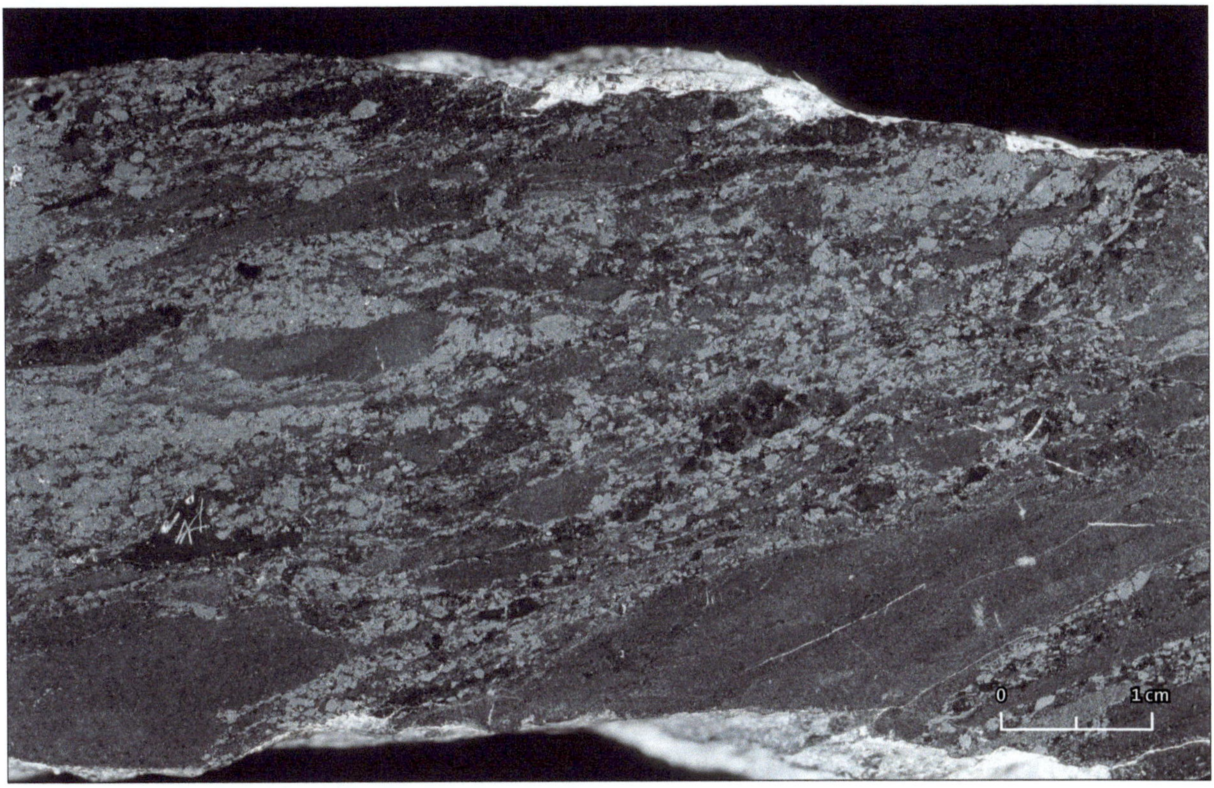

PLATE 20 Pronounced involvement of shear component (overprinted by successive hydrothermal events)
Starra (copper, gold) mine, Selwyn, Queensland, Australia.
(Specimen provided by N. Adshead-Bell).
(see also Plates 2, 3, 6, 7, 8, 9, 17, 18, 19,21,22, 34 and 35)

PLATE 21 Pronounced involvement of shear component (overprinted by successive hydrothermal events)
Starra (copper, gold) mine, Selwyn, Queensland, Australia.
(Specimen provided by P. J. Williams).

Plate 22 A B C D

Pronounced involvement of shear components (overprinted by successive hydrothermal events including brittle fracturing and probable hydraulic dilation). South Talbot gold mine, Kalgoorlie district, Western Australia, Australia.

The four illustrations opposite (*Plate 22 ABCD*) have been chosen to again illustrate the extremely common situation of a major ductile shear zone, overprinted by later more brittle fracturing associated with high pressure fluid dilation/brecciation.

The specimens illustrate a progressive sequence which commences at the outer perimeter of the shear zone (*Plate 22A*) moving into the central quartz-dominated vein zone (*Plate 22D*). The exact location of the gold is uncertain. The various components are best described with reference to all four plates.

Component 1 – The Main Early Stage Ductile Shear Zone

The host rock is a medium-grained equigranular igneous rock and although no fresh examples are visible it approximates to diorite. It is best seen via the paler grey zones towards the base of *Plate 22A* and the top of *Plate 22B*. The paler cream coloured materials were originally plagioclase feldspars, and the darker zones were ferromagnesian minerals (mostly amphiboles). Both major components now exhibit alignment (a ductile shear fabric). The feldspars are mildly sericitised whilst the ferromagnesian components are now varying combinations of "titano-ilmenite/leucoxene" and sericite? Fine-grained calcite is also present, and most of this alteration seems to be linked with the ductile shear development (early).

Component 2 – Dark Grey Quartz-Arsenopyrite ± Pyrite

This is well represented on all four plates. In *Plate 22A* it is weakly represented as a spaced crack style which both parallels and crosscuts the earlier shear fabric. These appear to be small infill veins (quartz – dark grey, arsenopyrite – pale silver). The veining is stronger in *Plate 22B* where it is clear that significant dark siliceous alteration accompanies the vein structures. Numerous large grains of arsenopyrite (silver) occur as isolated alteration spots (top right) possibly associated with the dark siliceous alteration. Arsenopyrite also occurs in other regions of the slab which on close insection can also be linked to minute dark impersistent (silica?) cracks. The dark "vein" structures mostly parallel the early rock fabric, although one angles across.

In the third plate (*Plate 22C*), the dark silica-arsenopyrite stage is even more strongly developed with the dark alteration replacing both feldspathic and ferromagnesian components components. Arsenopyrite alteration is similarly more extensive and large crystals and grains are easily visible, together with minor pyrite. The dark channelways are oriented parallel to initial rock foliation. Note the textural inheritance.

The final plate is a good illustration of the dark silica-arsenopyrite spotted stage, occurring as fragments in a mass of white quartz. Many of the fragments are partially altered to white-grey quartz.

Component 3 – White Quartz ± Arsenopyrite, Pyrite

The white quartz stage is dominated by infill and is clearly visible on all the plates. The relationship with the dark silica is a little questionable. The close spatial connection initially suggests that it is simply the infill stage to the dark silica -arsenopyrite. This is further supported by the presence of arsenopyrite (and pyrite) as coarse-grained infill (*Plate 22D* – top right). However there are two less convincing factors. Firstly the dark silica alteration is overprinted (altered) by the white silica (*Plate 22D*), and some of the white quartz veining is seriously asymmetric in relation to dark alteration (see *Plate 22B*). The white quartz is crystalline, not deformed, and although not visible in the plates, commonly contains small vugs (incomplete infill), and isolated prismatic crystals with a pearly lustre and suspected multiple twinning. They are probably albite.

Despite the questions concerning precise timing, it is evident that ductile shear structure has been a focus for more brittle style refracturing. It can be interpreted that the fluid responsible for white quartz is probably of the high pressure variety, being responsible for both vein dilation and dismemberment of wall rocks.

The identity of the yellow infill mineral in *Plate 22A* is unknown.

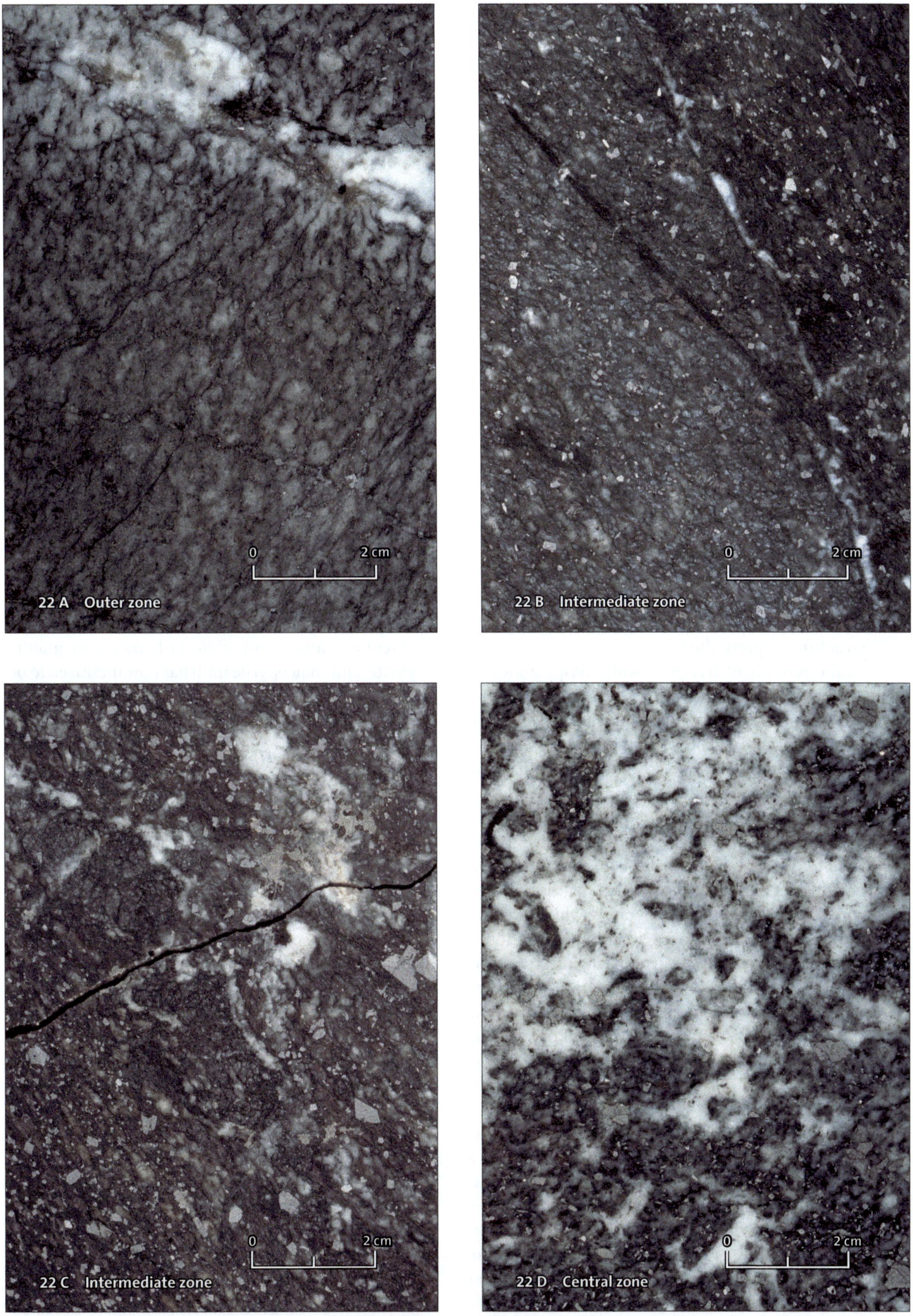

PLATE 22 **Pronounced involvement of shear components (overprinted by successive hydrothermal events including (brittle fracturing and probable hydraulic dilation)**
South Talbot gold mine, Kalgoorlie, Western Australia, Australia.

Plate 23 A B C D

Pronounced shear style system (multiple brittle style reactivations and multiple introduction of hydrothermal fluids). Dilation and hydrofracture via fluid pressure?
Ernest Henry (copper, gold) mine, Cloncurry district, Queensland, Australia.

Ernest Henry is located within a major regional ductile to brittle shear system which exhibits multiple reactivation and multistage introduction of hydrothermal fluids. The four examples illustrate repetitive breaking styles controlled by the presence of an original shear structure, local stress regimes within the wall rocks, and probable dilation due to fluid introduction.

PLATE 23 A

This section of core is composed of fine-grained brick red material which is a K-feldspar ± hematite altered meta-volcanic rock. The alteration is an early event which on a larger scale appears to relate to the broad scale shear structure, increasing in intensity as the structure is approached.

The prominent irregular white wispy zones are composed of carbonate (white) magnetite (dark) and minor chalcopyrite-pyrite (pale yellows). The zones are oriented and essentially represent infill with little or no alteration. The term "seagull texture" has been used as a local descriptive term.

The remaining dark spotted material is predominantly magnetite associated with fine-grained spots of sulphide (pyrite, chalcopyrite). Some biotite may also be present ± minor carbonate. This stage is actually a combination of infill and alteration which precedes the white carbonate stage. A careful inspection reveals that it is arranged as an oriented, interconnected meshwork, with an overall orientation paralleling that of the more obviously carbonate dominated stage. From a broken rock perspective, both oriented meshwork patterns are essentially cracks and create ovoid to elongate lensoid fragmentation. (See also *Plates 27–30*).

The oriented meshworks are crackle zones which are positioned at right angles to the main shear fabric, and give the impression that the peripheral crack zones, are largely controlled by pre-existing stress distribution in the wall rocks. The "seagull texture" results from the irregularly distributed dilation occurring during late stage fluid introduction, and it is suspected that the fluid pressure is responsible for the dilation. It is noted that this type of rock breaking is commonly picked out by later dilation and forms a good guide to locating nearby shear zones. It is tempting to use the term tension gash, although this is misleading if the opening is due to fluid pressure. A tension gash philosophy would involve a sense of movement.

PLATE 23 B

This plate is similar to *Plate 23A* but conveys the broken rock pattern in a more visual manner. The creation of irregular to rounded-ovoid-lensoid rock patterns is more obvious. The specimen is still within a marginal cracked rock zone, either adjacent to or between sheared zones (see *Plates 6, 7, 8 and 9*) but has been affected by greater fluid flow. In particular, the early dark stage is better developed and it is apparent that alteration associated with this stage is intense. The rocks have been altered via the development of fine-grained magnetite spotting and the brick-red feldspathically altered volcanic host rocks have deepened in colour.

PLATE 23 C

This core sample is from one of the "central" sectors of the fault system. The major difference is that the extent and amount of white carbonate ± magnetite, pyrite, and chalcopyrite infill has now increased to occupy some 40–50 % of the rock. The earlier wall rocks are now fragments, many of which assume irregular but generally rounded outlines. The rounding effect probably reflects the way the rock disintegrates (see *Plates 23A, 23B*) and has little to do with any milling effects. This interpretation is supported by the lack of any fine rock flour-sandy matrix. It would seem that the effect is due to the presence of a fluid under high pressure causing dilation/fragmentation. Any original structural information has been overprinted and effectively concealed by the successive feldspathic and magnetite alteration effects.

PLATE 23 D

This plate further illustrates the repetitive overprinting breaking theme. The rock comes from the fringe zone (crackle zone) of one of the major ore structures (see *Plate 23 A* above) and illustrates a further oriented crack style dislocation now represented by fine-grained pyrite. The previous red feldspathic ± hematite, dark (magnetite) and white (carbonate) stages are all represented.

The complete nature of the late pyrite-dominated stage is unclear. However, it is apparent that pyrite alteration plays a major role with pre-existing carbonate (calcite?) being replaced. Pyrite infill is strongly suspected, associated with the alteration. Note that all stages are similarly oriented.

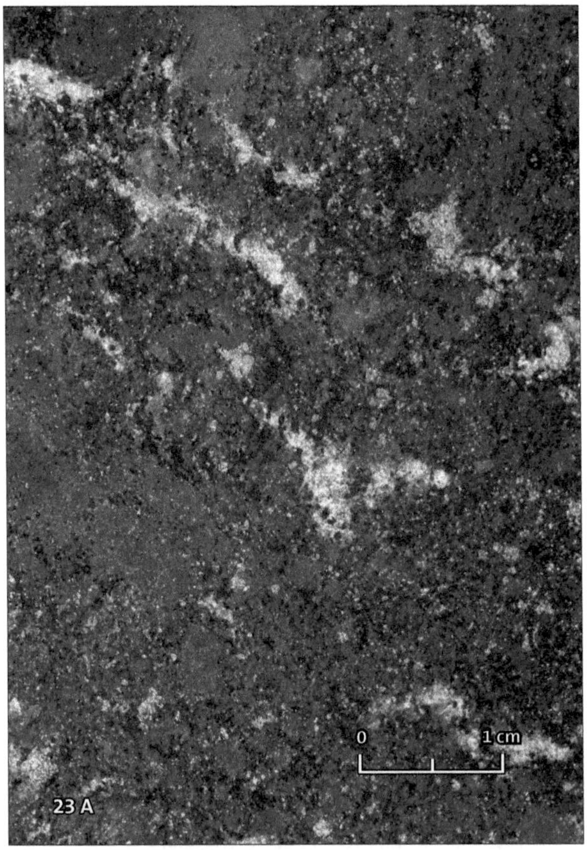

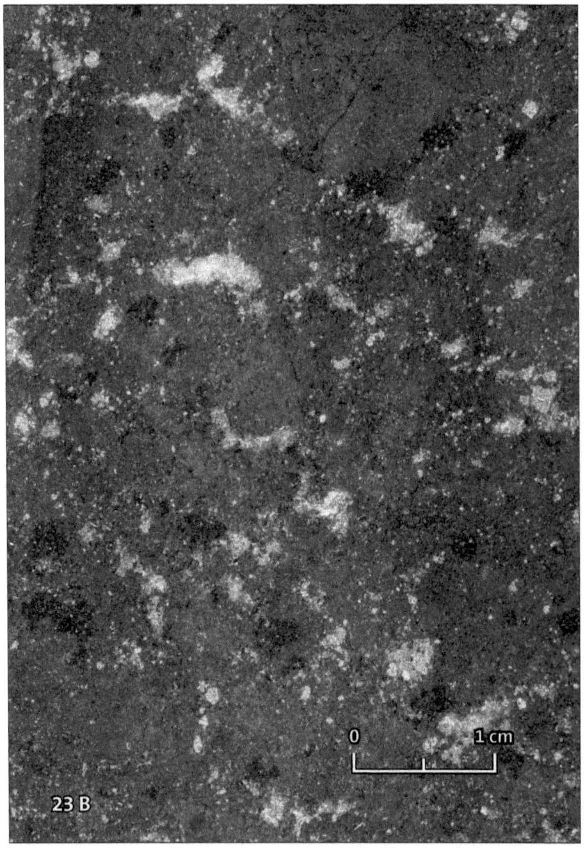

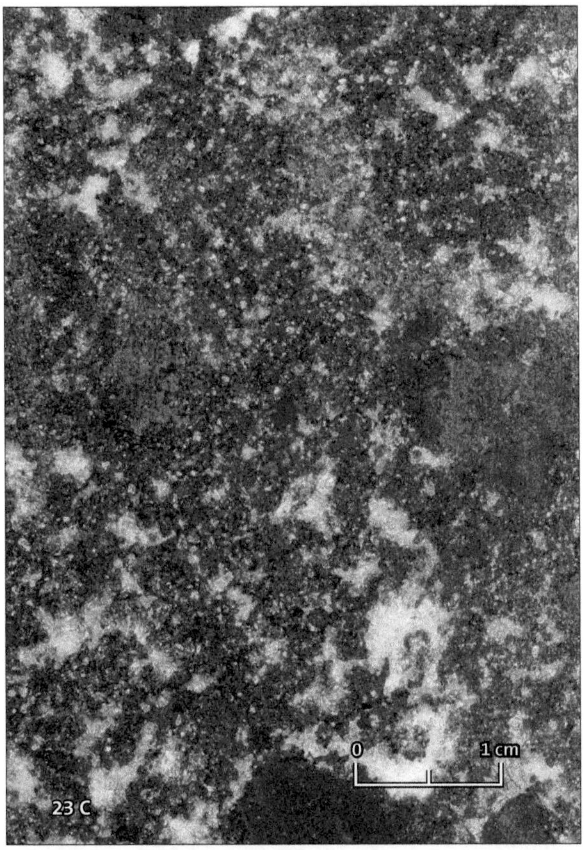

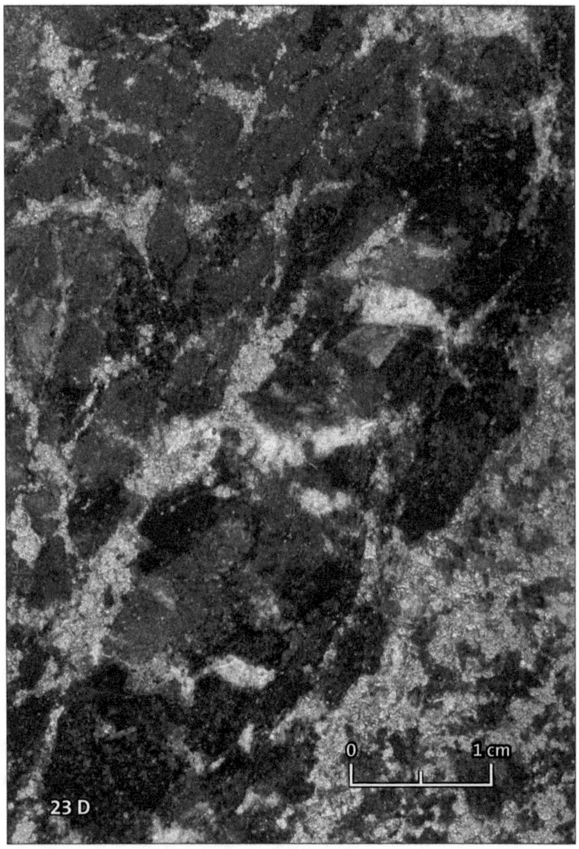

PLATES 23 **Pronounced shear style system (multiple brittle style reactivations and multiple introduction of hydrothermal fluids. Suspected dilation and hydrofracture via fluid pressure?)**
Ernest Henry (copper, gold) mine, Cloncurry district, Queensland, Australia.

Plates 24 A–F

Small scale examples of various brittle fracture styles

Plate 24 A

Stockwork
Red Dome gold mine, Chillagoe, Queensland, Australia.

A small scale example of shattered stockwork style fracture in fine-grained granite. The fractures are occupied by veins of quartz (grey) ± epidote (green) now overprinted by red-pink iron-oxides relating to weathering. This will be part of a larger system and it should be noted that once again the "stockwork" actually displays a strong degree of orientation within the major veins.

Plate 24 B C

Dilational Jog
Non Loc, Hidden Valley tin district, Queensland, Australia.

An excellent small scale example of two parallel fractures linked by shattered broken ground. The link mosaic pattern although brittle stockwork in style does exhibit a general orientation. This type of pattern is generally loosely referred to as a dilational jog and is very common at all scales within fracture systems. The linking fracture is considered to form via stress fields building up at the main fracture tips which are ultimately relieved via brittle semi-oriented rupture. The reason for stress concentration could be either directly tectonic (movement) or a fluid pressure build up related to tectonism. The "vein" system is composed of dark infill (cassiterite?) overprinted by iron-oxides. The white/pink patch also appears to be infill (feldspar?quartz?) and the grey background is a pre-existing siliceous? alteration of the fine-grained granite host.

Plate 24 D E

Cymoid Loop
Kangaroo Hills tinfield, Queensland, Australia.

A small scale example of a quartz vein system viewed on the weathered surface and on the sliced rock. The plate illustrates a species of breaking pattern, subsequently dilated via quartz infill (fluid pressure-precipitation) often referred to as a cymoid loop. Terms such as splay or even dilational jog might be employed. In reality it is just one section of a fracture system connecting up with another and there are many variations on the theme. The considerable variation in fracture dilation is normal and difficult to explain in any reasoned manner (fluid pressure variations versus local rock strengths?). This type of pinching and swelling is common in almost all vein systems at all scales.

Plate 24 F

Sheeted vein system
Waverley district, Kangaroo Hills tinfield, Queensland, Australia.

A small scale example of spaced parallel fractures with an infill of quartz (grey) and cassiterite (dark). This is a species of sheeted veining. The entire zone is only a few centimetres wide and it is a little difficult to envisage the style of stress/strain that would result in this style of fracturing. There seems to be little or no displacement involved. The material is locally called streaky ore. The granite host rock is partially albitised to the right, with potassium feldspar (pink) being replaced by albite (white). (See also *Plates 14 and 15*)

PLATE 24 Small scale examples of miscellaneous brittle fracture system (± probable hydraulic dilation) (All specimens provided by G. W. Clarke).

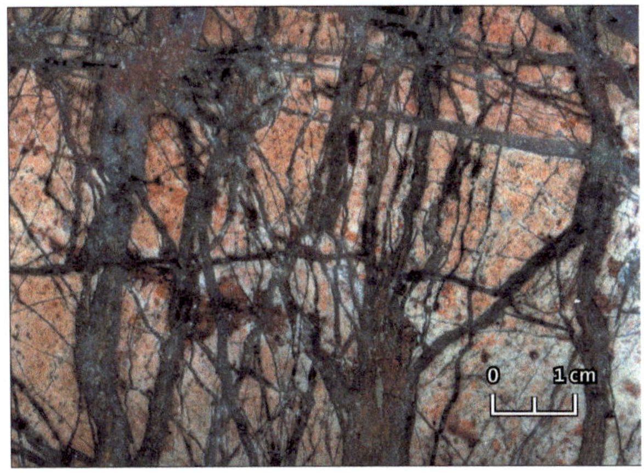

PLATE 24 A Stockwork Red Dome gold mine, Chillagoe, Queensland, Australia
(Small scale – see also Plates 31 and 32)

PLATE 24 B C Dilational Jog
Non Loc, Hidden Valley tin district, Queensland, Australia.

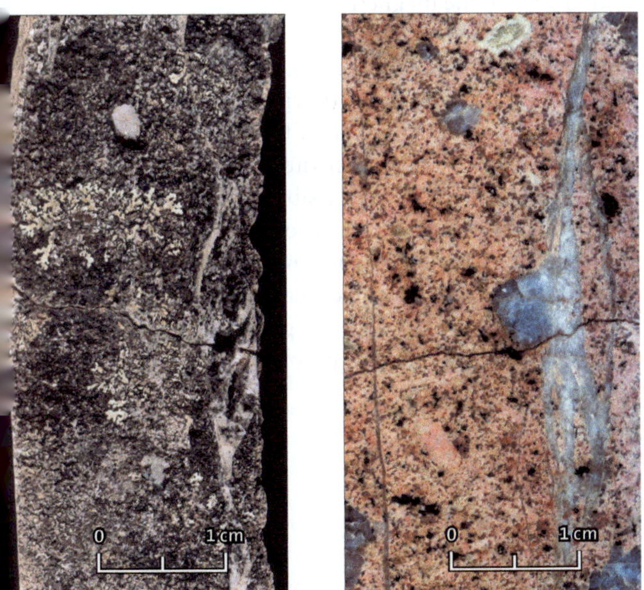

PLATE 24 D E Cymoid Loop
Non Loc, Kangaroo Hills Tinfield, Queensland, Australia.

PLATE 24 F Sheeted vein system
Non Loc, Waverley tin district, Kangaroo Hills Tinfield, Queensland, Australia.
Small scale – (see also Plates 14 and 15)

◘ Plate 25

Pronounced brittle fracture style (overprinted by hydrothermal alteration) Non Loc, Kangaroo Hills tinfield, Queensland, Australia.

This specimen is an excellent example illustrating an extremely common brittle style fault situation and a range of attendant breaking patterns.

A pre-existing quartz vein/silicified rock (white-grey) has been a focus for renewed fault movement. At the bottom of the plate very little movement has occurred, and the rock has broken in an anastomosing pattern creating a variety of shapes including lensoid and simple crack style fracture.

The central and upper regions have clearly suffered more extensive disruption, creating both angular and semi-rounded fragments which range from equant to elongate. Most of these shapes can readily be seen in the cracked materials within the lower region of the plate. This general format is sometimes referred to as cataclastic (brittle, with lots of angular fragments in a finer gouge).

The broken rock with a gritty quartzose matrix forms an extremely permeable unit which in this case has channelled hydrothermal fluid. The dark material is tourmaline ± quartz, and is present as partial to complete alteration of the fine-grained matrix and larger fragments. Very little infill is discernable and a low pressure fluid infiltration is suggested.

The top right of the plate reveals a relatively unaltered zone with a more visible matrix. This demonstrates that fragmentation (faulting) preceded fluid introduction. This is the usual situation, i.e. fracturing precedes hydrothermal fluid circulation.

A more recent crack pattern overprints the rock as a minute pale network of infill veining plus broken rock. (See also ***Plates 10, 11, and 13***).

◘ Plate 26

Multiple brittle fracture system (extensively overprinted and obscured by successive stages of alteration/infill) Candelaria (copper, gold) mine, Copiapo district, Chile.

This plate illustrates the normal situation revealed by any close examination of mineralisation, with successive stages of fracturing/faulting being utilised by successive stages of hydrothermal fluid introduction. The latter produce infill/alteration which commonly conceals the situation from casual viewing. In outcrop the rock is extremely dark, and reveals little of the extensive multi-fracturing history. The specimen (***Plate 26***) has been slabbed and ground flat, and several stages are now apparent.

The initial stage within this rock is magnetite (black) which occurs as a total fine-grained alteration of pre-existing rocks. Elsewhere it is apparent that this too is brittle fracture controlled, and is preceded by a similarly structurally controlled biotite alteration stage. The original wall rocks are mostly andesitic.

The brittle magnetite has been fractured, and succeeded by a fluid which precipitated fine-grained crystalline prismatic green amphibole and white calcite. This is clearly infill and thus implies a reasonably high fluid pressure. It is probable that some hydrofracturing is involved and undoubtedly the patterns observed here are part of a much larger system (Candelaria is a major copper deposit!). Note the lensoid fragmentation style of the magnetite.

The final stage is essentially controlled by an overprinting brittle crackle system, now marked by the yellow sulphides (pyrite and chalcopyrite). These are a combination of infill and alteration. Infill predominates in the magnetite host where excellent trianagular textures are visible. It is possible that some of the carbonate has been altered to sulphide. The style of the infill component suggests a mild fluid pressure may have assisted dilation.

The small rock is actually part of a much larger system (scale perception problem – see ***Plates 1, 2 and 7***).

PLATE 25 Pronounced brittle fracture style (overprinted by hydrothermal alteration)
Kangaroo Hills tinfield, Queensland, Australia.
(Specimen provided by G. W. Clarke).

PLATE 26 Multiple brittle fracture system (extensively overprinted and obscured by successive stages of alteration/infill)
Candelaria (copper, gold) mine, Copiapo district, Chile.
(Specimen provided by P. J. Pollard).

Plate 27

Brittle style rock fracture pattern with rounded to ovoid fragmentation of garnet host rocks (skarn)
Wabu gold prospect, Bilogai, Papua, Indonesia.

The main reason for inclusion of the plate opposite is to demonstrate that contrary to popular conception there are many rocks which actually break up naturally into ovoid and lensoid shapes. (See also *Plate 23*).

The dark ovoid shapes are composed of myriads of pinhead size granular garnets. They were originally pale green (mid-top right) but are now substantively altered (dark green-chlorite?).

The garnet grain size suggests that the original rock was fine-grained and the local context indicates a host rock of fine-grained carbonate. This alteration (skarn) commonly results in a garnet-carbonate assemblage, although no carbonate is present here.

The paler network is also garnet (pale pink-cream) which is crystalline and more variable in crystal size. The fracture pattern is interpreted as a crackle style, subsequently invaded by a pressurised hydrothermal fluid which dilated the cracks and precipitated garnet.

The pink garnet infill is commonly incomplete and has created a very porous/permeable rock. The remaining interstitial spaces have been variably infilled by later stages, including very dark chlorite, grey calcite, and rare sulphides (pyrite, sphalerite? hematite?).

It is probable that the chloritic alteration of the green garnet relates to one or more of these.

Plate 28

Brittle style rock fracture pattern with two generations of rounded to ovoid fragmentation of garnet host rocks (skarn)
Wabu gold prospect, Bilogai, Papua, Indonesia.

The skarn assemblage and textures opposite should be compared with the illustration above (*Plate 27*).

The main feature to note is that not all rocks break up into angular fragments. The host rocks here are already fragmental, and although difficult to discern are actually the pale green and later cream-pink garnets depicted above (*Plate 27*)

The pale green alteration garnet is well-illustrated at mid-top left and is substantially altered to carbonate, chlorite(?) and probably epidote. The subsequent infill of pink garnet is visible at many points, particularly to the edges of the finer green-grey clumps, where the spotty nature actually reflects coarser-grained zoned garnet crystals. The rock dislocation naturally occurs along the pre-existing coarse-grained garnet infill channelways (see above) and results in fragments with variable amounts of coarse-grained garnet attached, isolated small granular clumps and even crystals of pink garnet.

Coarse-grained white to pale pink carbonate is also fragmented, and seems to have originally infilled post pink garnet (top-mid-right).

The prominent dark grey material is infill magnetite ± dark grey carbonate, which is interpreted to result from precipitation from a high pressure fluid. This has gained entry via a fracture system and then proceeded to dismember the granular rocks. The magnetite does not cause any significant alteration of the garnet wall rocks. A few sulphides (mostly pyrite-yellow) occur within the magnetite and it is not clear whether these are late infill or alteration of carbonate.

The zone is part of a much larger, layer parallel fault system, some metres above at a carbonate/metasediment boundary.

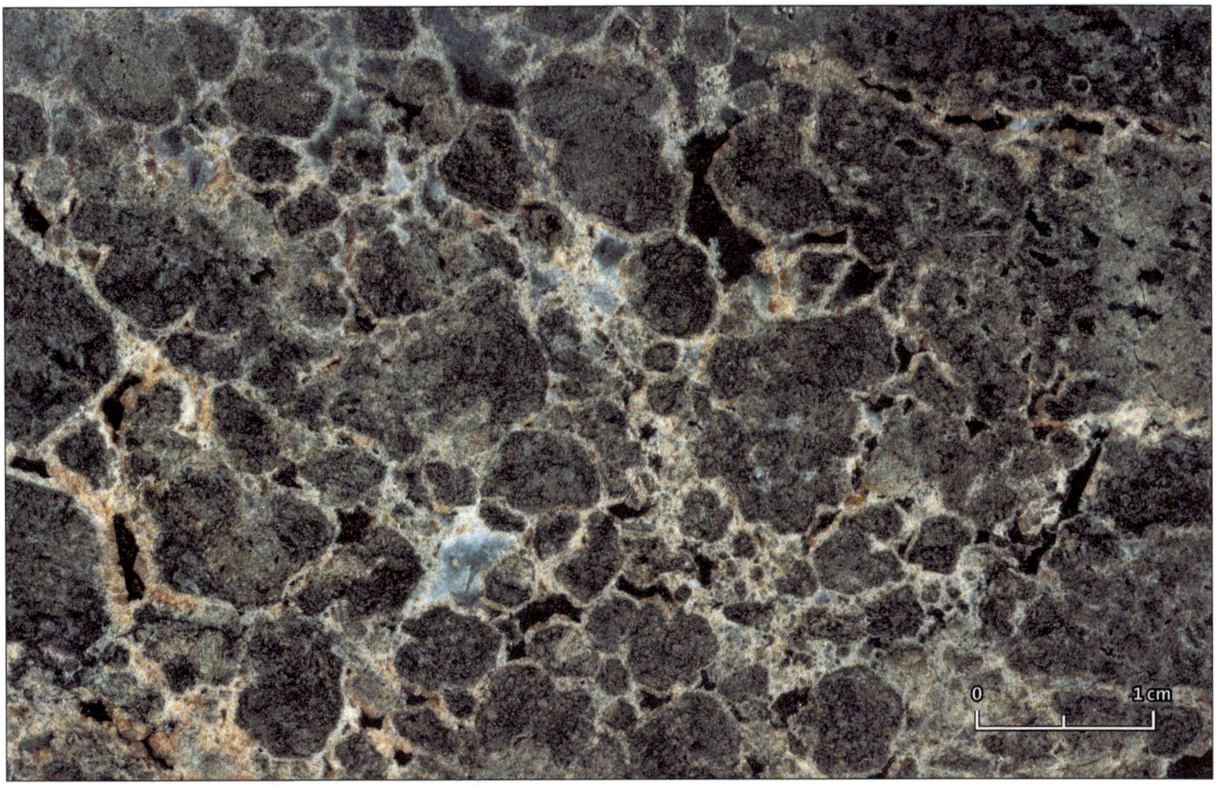

PLATE 27 **Brittle style rock fracture pattern with rounded to ovoid fragmentation of garnet host rocks (skarn). Wabu gold prospect, Bilogai, Papua, Indonesia.**

PLATE 28 **Brittle style rock fracture pattern with two generations of rounded to ovoid fragmentation of garnet host rocks (skarn) Wabu gold prospect, Bilogai, Papua, Indonesia.**

Plate 29

Brittle style rock fracture patterns (concealed by multiple alteration/infill overprinting, rounded to ovoid fragmentation)
Cannington (silver, lead, zinc) mine, Cannington, Queensland, Australia.

The plate opposite has been included to illustrate just how difficult it is to observe broken rock textures within multiply broken, highly mineralised ore systems. A good understanding of infill, alteration and breccia textures is critical, and obtaining a good view via cut slabs is almost mandatory. The second major point of the photograph is to illustrate that, contrary to intuition, there are many situations where host rocks fragment in a rounded/ovoid manner. It is always a good idea to move away from the intensely broken areas and look at the crackle patterns towards to perimeter. These show how the rock will ultimately break up.

The plate illustrates that a pale green host rock occurs as rounded, almost circular fragments (see also *Plates 23, 27 and 28*). This rock is somewhat unusual, resulting from alteration/infill of sillimanite grade metamorphic rocks (schist, gneiss, amphibolite). The extensive alteration overprint results in a pyroxene (hedenbergite)-pyroxenoid (pyroxmangite) – amphibole (hornblende) assemblage. The resultant green rock is very granular and ultimately controls the ovoid breakup pattern (see also garnets, *Plates 27 and 28*).

The granular pyroxene assemblage is cracked and broken into rounded shapes in association with a second hydrothermal fluid which has produced magnetite as infill and alteration of the pyroxene assemblage. This is difficult to see on the plate, but is displayed (mid section of basal edge) where pale grey materials (magnetite) can be seen surrounding pale green fragments (pyroxene). This occurs inside the obvious fragments.

Two more stages are readily apparent although the relationship between them is obscure.

A major galena stage (silver coloured) is very obvious, and is clearly dominated by infill textures. At microscopic level, minute needles of Mn-rich chlorite (pyrosmalite) partially line some of the cavities. The galena is accompanied by a wide range of silver-rich sulphosalts, and from the infill nature it can be interpreted that a reasonably high pressure fluid has dismembered the previous wall rocks via a fracture network, presumably involving some hydrofracking.

The other stage is defined by the darker green zones which are also soft and granular (slightly plucked out via polishing). Chlorite and a pale green-buff vitreous component (fluorite?) predominate with a rare brown red vitreous equant mineral. Under a high power lens (15 ×) the galena stage appears to cut these components, which in turn cut the magnetite. The dark green stage is again predominately infill. Obviously, some microscopic detailing would sharpen mineral identification, but the rock seems to have suffered at least four stages of brittle fracturing (± effects of high pressure fluids).

Plate 30

Brittle style rock fracture (rounded fragmentation with probable hydraulic dilation)
Cannington (silver, lead, zinc) mine, Cannington, Queensland, Australia.

The second plate from the Cannington silver-lead-zinc deposit (see also *Plate 29*) also demonstrates the propensity for the altered wall rocks to fragment in ovoid patterns. The fragments are composed predominately of granular pyroxene (green) and quartz (grey) and are set in a matrix of coarse-grained sphalerite (brown) galena (silver) and rare creamy fluorite? The latter are undeformed and from microscopic examination appear to be infill. Minor sphalerite alteration may be present as small brown spots within some of the fragments. The recognition of infill immediately implies that at one stage the majority of this rock was a fluid filled cavity and it is presumed that the fluid was under high pressure, which probably assisted rock fracture/dismemberment. This pattern is identical to that in *Plate 27*. It also makes the point that not all fragment rounding is necessarily related to milled style (gas release) brecciation. The rocks simply break into ovoids.
(See also *Plate 23 C*).

PLATE 29 Brittle style rock fracture (concealed by multiple alteration/infill overprinting, rounded to ovoid fragmentation) Cannington (silver, lead, zinc) mine, Cannington, Queensland, Australia.

PLATE 30 Brittle style rock fracture (rounded fragmentation with probable hydraulic dilation) Cannington (silver, lead, zinc) mine, Cannington, Queensland, Australia.

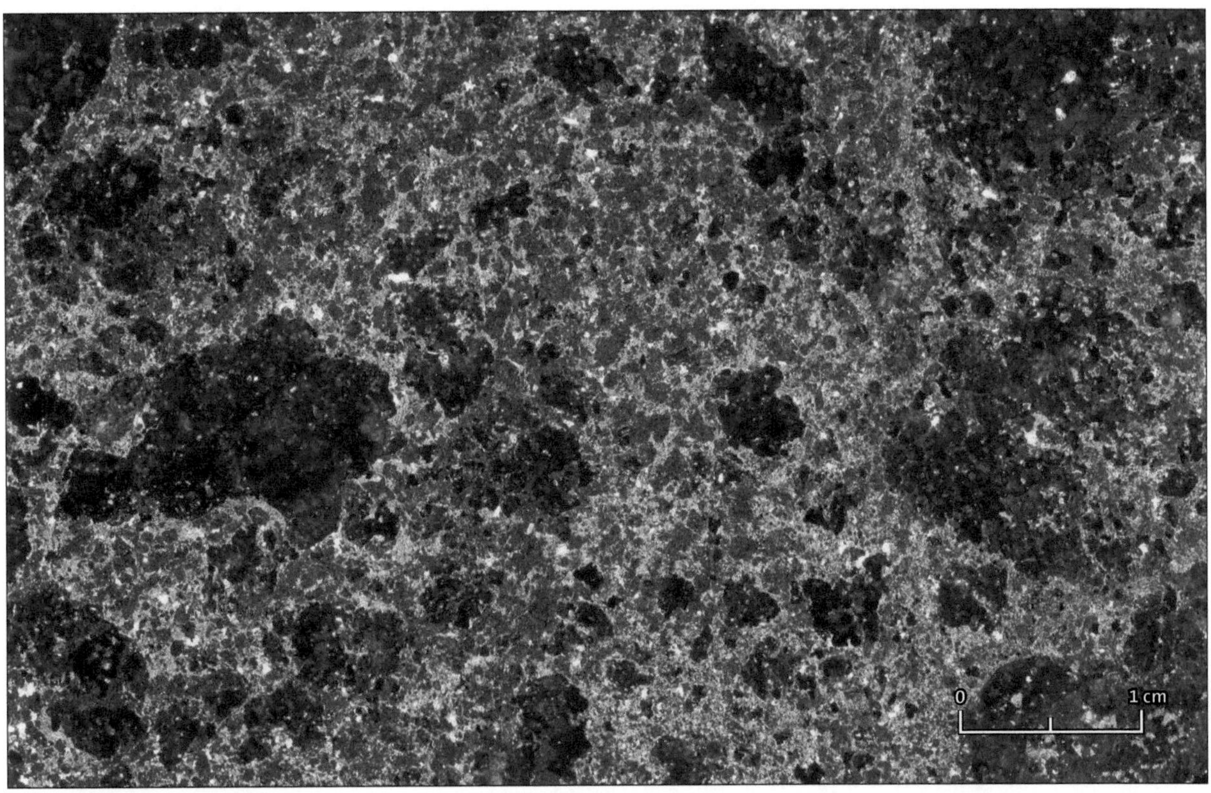

◻ Plate 31

Brittle style rock fracture (multiple stockwork)
Grasberg (copper, gold) mine, Papua, Indonesia.

The plate opposite has been selected to illustrate multiple brittle stockwork style fracturing, which typifies most of the porphyry copper style deposits on a global scale. The fracturing commonly develops at kilometre scale, and the reasons for the extensive repetitive style are a little obscure. Commonly there is little or no obvious connection with significant faulting, although the associated intrusive rocks were initially emplaced via large scale fracture controls. The style seems related to repeated focussed pressure from directly below.

This specimen is a small scale example from the Grasberg porphyry copper-gold system where a multiple stockwork system some 1 km × 600 m in surface area extends for over 1.5 km in depth. The system is complex but contains an early stage barren quartz stockwork which increases in intensity towards a central intrusive unit, located within a kilometre scale diatreme.

The quartz stockwork is depicted in the opposite plate as a set of grey-blue quartz veins which are composed of infill crystalline quartz, suggestive of dilation via high pressure fluid activity.

The quartz stockwork is in turn overprinted by millimetre scale stockwork style brittle fracture veins containing coarse-grained black infill magnetite. This relationship is not easy to see on most of the plate. However careful inspection will reveal dark veins along quartz vein margins (top left – centre right) and also dark veins vaguely central to quartz veins (centre left and right).

The sulphides (yellow) represent a further brittle stockwork fracturing and are present as infill (chalcopyrite ± bornite, pyrite, gold and molybdenite). The fracturing crosscuts the quartz veins and also channels along their margins and central zones after incomplete quartz infill (magnetite occurs similarly). The sulphide stage accounts of 99 % of the copper grades at Grasberg, and is the virtually the last stage in a long history of fracturing and fluid circulation. Minor gypsum overprints the sulphides.

The host diorite has been totally replaced by a major fine-grained magnetite alteration (dark background) which elsewhere can also be seen to be brittle stockwork controlled. In the rock opposite this has been superimposed on a major potassium feldspar alteration (pale tinges in right hand corner). No fresh rock exists on the illustration!

The major iron content (magnetite) raises the question of whether or not the occurrence should classify as an iron-oxide-copper-gold style deposit as opposed to a more conventional porphyry copper system?

PLATE 31 Brittle style rock fracture (multiple stockwork). Grasberg (copper, gold) mine, Papua, Indonesia.

Plate 32

Brittle style rock fracturing (multiple stockwork)
Grasberg (copper, gold) mine, Papua, Indonesia.

This plate is an excellent example of repetitive brittle style cracking, with successive stages of hydrothermal fluid introduction and mineral deposition/alteration.

The specimen illustrates some of the complexity within the Grasberg porphyry copper-gold system, and a style of rock breaking which is extremely common in the mineralised porphyry environment.

Surprisingly, this multiple stockworking style has not been well-documented, and there are very few accounts concerning the detailed distribution and variations within the individual stages over an entire mineralised occurrence. Consequently the mechanisms of large scale rock fracturing seen in porphyry domains are not well-understood.

This particular specimen is complex, but at least four stages can be seen:

1. Late cracking associated with small scale infill of sulphides (yellow). This is the main mineralisation stage and is actually variably composed of chalcopyrite, bornite, pyrite, gold with minor quartz, molybdenite and hematite.
2. Grey-white quartz veins overprinted by the above sulphide fracture network. The quartz veins are in a subparallel E-W orientation with linking fracture, and seem to have dilated in response to high fluid pressures? Hydrofracture may have assisted development of this stockwork.
3. Magnetite veins (dark coloured) occurring as a fine crackle network underlying the main quartz veins. This network is extensive and extends down to very small crack styles which interlink with larger examples. Nearly every black spot visible in the rock is actually magnetite which occurs as an alteration product associated with the vein infill. Pre-existing mafic minerals are favoured but not exclusive magnetite replacement sites.
4. Feldspathic alteration, which occurs underlying all the crackle domains as diffuse white/cream zones of major potassium feldspar alteration. The strong alteration component conceals most of the original channelways, but hints of small scale fracture control are evident in many places.

The greenish grey tones (centre-left) are probably the least altered rock, but still exhibit blurred igneous textures (sericitised?)

PLATE 32 **Brittle style rock fracturing (multiple stockwork). Grasberg (copper, gold) mine, Papua, Indonesia.**

Plate 33 A B

Brittle style rock fracturing, probably deep seated related to a shear system (dilation probably by fluid pressure-typical slate-belt style)
Non Loc, Hodgkinson goldfield, Queensland, Australia.

The fracture pattern displayed in the adjacent plates has several points of interest, and represents a common theme in gold-related quartz vein domains associated with slate and flysch style environments (see also *Plates 34 and 35*).

The wall rocks here are composed of fine-grained, dark argillaceous zones interlayered with fine-grained, paler more siliceous domains. The metasediments were probably shales/mudstones originally. The lower half of the rock (grey) is a more massive arenaceous unit and the entire sequence has been substantively deformed. Small scale fold development is well-displayed in the upper interlayered sector.

The lower more massive unit exhibits several examples of early pale grey/white short length quartz veins, possibly resulting from localised brittle fracturing within a more competent domain (see ***Plate 18A, 18B***).

The most obvious feature is the spaced array of sub-vertical white quartz veins. The quartz appears massive, but on close inspection exhibits many hints of coarse crystallinity and excellent crystal shapes are visible at higher magnification within the zones which contain pale green ± dark spots. They are just visible in the close up plate (top right, ***Plate 33B***). This whole quartz style is very typical of deep-seated quartz veins which form initially via coarse crystal growth, but exhibit stress-related recrystallisation due to superimposed local tectonic and/or deep-seated general rock stresses. This blurs the crystallinity, and not surprisingly the style predominates in deeply penetrating structures (faults/shears) – slate belt and greenstone gold related environments.

The environment is also a favoured home for fibre vein style quartz although none is visibly present here. The width of the infill suggests that extremely high fluid pressures are probably responsible for the dilation. Again this is typical within this domain style. The nature of the original rock fracture is obscure. However, the lack of shear fabric and/or broken rock associated with the quartz indicates a crack style. The layering seems to match across the veins and little or no movement is suggested. However, it is probable that a scale problem is involved and that the veins belong to a much bigger system involving both shear and brittle (crack) components.

It should also be noted that the quartz on the left hand side (***Plate 33A***) exhibits some extremely irregular patterns, with blobs, bends, bulbs and many other shapes. Once again, this is typical of this style and is interpreted to represent the result of extremely high pressure fluids which gain initial access via faults/fractures and then exploit any other scale of structural weakness (e.g. layering, cleavage, individual mineral partings) to shoulder rocks aside at the small scale. This commonly results in veins/structures that look deformed. This mechanism of high pressure fluid/fracture exploitation is probably similar to that of magma intrusion, and it is interesting to conduct a change of scale exercise, and view all the quartz as granite. The veins become stocks (say 0.25–0.5 km scale) and the irregular shapes become other forms of intrusive (magma chamber) shapes?

The factors creating high pressure can only be surmised, but it is probably no accident that many studies envisage prolonged tectonic processes (deformation) and development of this vein style as vaguely synchronous. Any hydrothermal fluid could easily develop very high pressure when caught up in fault movement and general deformation processes. Veins of this style commonly contain gold, although a close study almost always reveals that the gold was introduced much later as an overprinting stage which followed and refractured the regional scale quartz veins (***Plate 34***).

A close examination of the veins (***Plate 33A***) indicates several crosscutting crack zones linking to the pale green zones. A good example is present on the close-up (***Plate 33B***). These are composed of minute infill assemblages including grey quartz, pale yellow pyrite, darker crystals of sphalerite and galena(?), green prismatic/acicular chlorite(?), and minute pale coloured specs (muscovite?). The wall rock quartz is commonly darker or greenish tinged (alteration). Within some of the larger zones (top to mid left ***Plate 33A***) infilled vugs are present, and it seems that the initial white quartz stage failed to completely fill the vein structure. The pyrite also occurs as an alteration component, forming isolated yellow spots which commonly favour the dark argillaceous layers of the rock. It is possible that these originally argillaceous layers have been thinned via pressure solution, and now represent residual carbon concentrations. These are natural sites for sulphide and gold development (oxidation/reduction reactions).

PLATE 33 A Brittle style rock fracturing, probably related to a shear system (dilation probably related to fluid pressure typical slate-belt style)
Non Loc, Hodgkinson goldfield, Queensland, Australia.

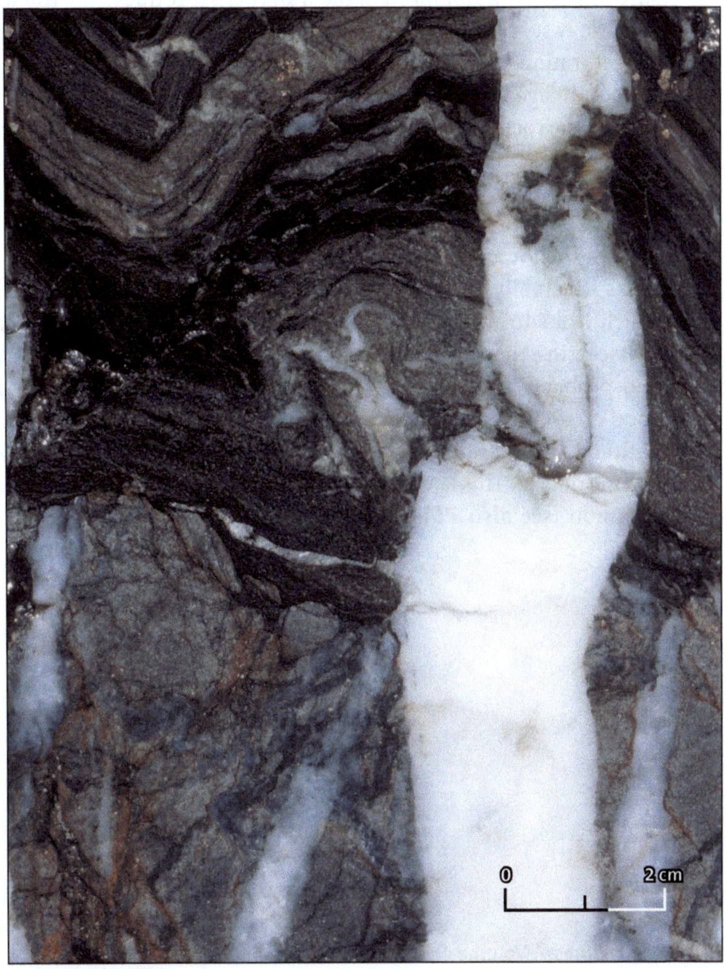

PLATE 33 B Brittle style rock fracturing, probably related to a shear system (dilation probably related to fluid pressure typical slate-belt style)
Non Loc, Hodgkinson goldfield, Queensland, Australia.

4.4 Textures of Related/Miscellaneous Interest

◻ Plate 34

Post depositional shear strain (dissolution, stylolite development, boudinage, flattening and recrystallisation) Stawell gold mine, Victoria, Australia.

This specimen has been included under slightly false pretences in that very little of the texture would merit the title of broken rock in conventional usage. However, the rock is fairly typical of quartz veins contained within argillaceous metasediments-slates (see *Plates 33A, 33B*) which have been subsequently subjected to extremely high pressure shear strain. The black wavy zones are now composed dominantly of carbonaceous materials, but were probably originally semi-parallel rafts of meta-sediments occurring between and within a set of closely spaced quartz veins. Severe compression (shear strain) resulted in "pressure solution" (dissolution) of the wall rocks, producing a wavy stylolitisation effect with a residual carbon concentration. The more competent quartz veins tend to boundinage, flatten, develop bulges and even minor folds as the rock accommodates the strain via ductile response. Technically the process should be referred to as shear controlled dissolution.

Although not well displayed on this plate, much of the paler yellow colour within the carbonaceous wall rocks is actually layer controlled suphide alteration which links to minute sulphide (pyrite rich) veins crosscutting the quartz. This pyrite is undeformed and probably represents the gold-bearing stage. A good crosscutting example can be seen towards the bottom left hand side as a low angle vein clearly crosscutting quartz. This late brittle overprint is extremely common (see also *Plates 33A, 33B*) and sulphide alteration within the carbonaceous layers is popularly considered to be via a reduction reaction.

The more visible pale buff-orange throughout the plate is an iron-oxide weathering product related to pyrite oxidation via meteoric fluids. A dextral sense of shear is well displayed by the central quartz ovoid/lozenge. The original coarse-grained infill quartz has been recrystallised during deformation. Fibre quartz vein development is common in this domain but not depicted here.

◻ Plate 35

Post-depositional shear strain (dissolution, stylolite development, boundinage, flattening, recrystallisation) Non Loc, Victoria?, Australia.

The lower plate illustrates the ultimate extension of the deformation fabric seen in (*Plate 34*).

The anastomosing semi-stylolitic, dark graphitic bands (pressure solved metasediments) and the boundinaged, distorted, milky recrystallised quartz testify to considerable shear strain and layer shortening. It is quite possible that this rock originated as a set of high pressure fluid style quartz veins within a shear zone. Subsequent shearing compression and shear related pressure solution has resulted in the current format. Minor pyrite grains occur within the carbon-rich seams and minor, late crosscutting quartz is visible. Rocks of this nature are commonly termed ribbon or laminated veins. They are common in slate belt, flysch and greenstone gold environments (see *Plates 33A, 33B*).

PLATE 34 Post-depositional. Shear strain (dissolution, stylolite development, boundinage, flattening and recrystallisation) Stawell gold mine. Victoria. Australia.

PLATE 35 Post-depositional. Shear strain (dissolution, stylolite development, boudinage, flattening and recrystallisation Non loc, Victoria?, Australia.

◻ Plate 36

"Saddle Reef" (variations)
Cosmo Howley gold mine, Northern Territory, Australia.

The plate opposite illustrates an unusual(?) form of rock "breaking" with the result resembling a saddle reef style of ore deposition. However, careful examination reveals no signs of associated shearing which commonly accompanies the classic saddle reefs of slate belt gold fields.

The rock here is a fine-grained carbonaceous shale/slate which has been tightly folded, and probably contains an approximately layer parallel early cleavage (not visible). Specific zones of high strain are visible (left side) which represent a differentiated crenulation cleavage.

The sulphide (yellow, white ± quartz) is fine-grained pyrite and clearly represents infill, with some open vugs (dark) revealing minute pyrite crystals. The finer grained pyrite sporadically developed as blebs and spots along layering is sulphide alteration, and readers should note that this texture is extremely characteristic of layer controlled sulphide alteration. The sulphides are typically elongate and discontinuous in detail.

The major question here concerns the development of the cavities. These clearly favour the general horizontal direction, and to some extent the fluids do not seem to be exploiting the vertical high strain zones.

The favoured interpretation is that high pressure fluids were introduced laterally from a nearby major structure. The fluids essentially travelled horizontally (i. e. from behind or in front of the photograph) and sought out weak points to cause inhomogeneous dilation. In short, the layers were broken apart selectively, with a preference for hinge zones.

An alternative suggestion is that the "open space" was formed via deformation, with subsequent fluid introduction. It is difficult to supply definitive evidence for either process, and both could be involved.

PLATE 36 "Saddle Reef" (variations)
Cosmo Howley gold mine, Northern Territory, Australia.
Specimen provided by W. P. Laing).

5 References

Laznicka, P., 1988. Breccias and coarse fragmentites: petrology, environments, associations, ores. Amsterdam, Elsevier, 832 p.

Volume 5 – Broken Rocks · Breccia II

Volume Front Page

Specimen from the Mungana (intrusive) breccia pipe, Chillagoe, Queensland, Australia. The specimen has been selected primarily because it "looks nice" and forms a spectacular cover illustration. It is very oxidised and displays a wide range of different argillised-, silicified- and iron-oxide-altered fragments within a prominent fine-grained (iron-stained) matrix. The fragments are a mix of angular and rounded, with the milling being more apparent at the smaller scales. The pipe is associated with gold mineralisation and interpreted as a diatreme style (*Broadhurst, 1953*).
Specimen provided by G. W. Clarke.

1 Initial Recognition

2 General Features – Range in Styles and Scale of Intrusive Breccia

Introduction, Scope and Arrangement of Volume

This volume is designed to illustrate textures and field identification problems associated with a group of fragmented rocks which are loosely termed "intrusive breccia". The intrusive terminology is derived from their tendency to contain numerous rounded, milled fragments which exhibit evidence of upward transport and forceful injection to form pipe and branching dyke-like structures. The breccias are commonly hosts to overprinting economic mineralisation and with their tendency to reach maximum development within well-mineralised epithermal and porphyry environments, are of major exploration interest.

The term "intrusive breccia" has been selected by the author as an appropriate general family name for the purposes of textural illustration, however, readers should also be aware that within the jungles of breccia nomenclature there are numerous other terms with descriptive and genetic overtone which are utilized within this general grouping. These include milled, explosive, explosion, cobble, volcanic, diatreme, hydrothermal, hydrothermal intrusive, magmatic-hydrothermal, phreatic, phreatomagmatic, hydromagmatic, pebble dyke, tuffisite and fluidised. The author would also confess to having added to the confusion by suggesting to generations of students that the field term "push-up breccia" is a useful catch-all phrase.

It is not proposed to become enmeshed in the problems of either terminology or classification systems within this small volume. The latter has defeated generations of breccia specialists, and readers are best referred to an especially gallant attempt by *Sillitoe (1985)* to place a volcanoplutonic arc perspective upon the many variations of this (and other) breccia styles. It should be noted that from a field perspective he regretfully concludes "a continuum exists between many of the breccia types and it is difficult to identify unique criteria for their distinction" (*Sillitoe, 1985*). In terms of genesis, a major source of energy is clearly required to comminute rock fragments and arrange them into pipe- and dyke-like formats. There is a spread of opinion concerning magmatic versus meteoric fluid input, but most authors are in general agreement that the mechanism involves sudden, localised, potentially violent explosive release of expanding high-pressure fluids. Common suggestions embrace.

1. Decompression mechanisms and related phase separations within:
   (a) magma
   (b) magmatic hydrothermal fluid
   (c) meteoric fluid
2. Second boiling creating phase separation within naturally crystallising magma
3. Boiling of meteoric water and consequent phase separation:
   (a) Via heat transfer from a nearby magma chamber to adjacent groundwater
   (b) Via ground water and magma coming into direct contact.

Interested readers are referred to summary overviews {*Sillitoe (1985), Allman-Ward et al. (1982), Burnham (1985), Hedenquist and Henley (1985) and Sheridan and Wokletz (1983)*}.
After considerable thought the text has been arranged into six sections, including:

1. *Initial recognition.* This section aims to introduce the reader to the very common problem called *total lack of recognition*. It commences with some easily recognised situations and progresses to increasingly obscure examples.
2. *Ranges in styles and scale of intrusive breccia.* Intrusive breccias range from extremely large breccia pipes at kilometre scales, to narrow sections of milled materials within epithermal veins. Some concepts of style and scale changes are illustrated.
3. *Multiple overprinting and/or textural modification.* Intrusive breccia forms a natural focus for hydrothermal fluid flow and consequently hydrothermal alteration. The breccia sites are usually involved with repetitive fracture zones, and multiple overprinting commonly makes identification and paragenetic studies very difficult. Several examples are illustrated to demonstrate recognition and procedural problems.
4. *Selected special features of intrusive breccia.* This section is designed to show some selected features which may be encountered within individual examples of intrusive breccia.

5. *Intrusive breccia associated with maar-volcanoes and associated diatreme structures.* Numerous epithermal (and porphyry) systems can be related to intrusive breccia diatreme structures, and this section is designed to provide the reader with some overview of the maar-volcano environment.
6. *Intrusive breccia in other epithermal domains.* A short section to illustrate some further epithermal styles.

1 Initial Recognition

It may come as a surprise to many readers who are new to the breccia world to learn that one of the biggest problems is *total lack of breccia recognition*. The non-recognition problem is particularly true in relation to "intrusive breccia" styles. It would be possible to document some ten to twenty examples here where "total lack of recognition" has proved to be not only geologically embarrassing but also extremely costly. There is perhaps little point in specific "finger pointing", but this is a very real and potentially serious problem. The first section of this volume attempts to illustrate easy versus difficult to recognise intrusive breccia exposures, and some case history lessons are embodied in the plate descriptions. The factors working against recognition are listed below and several of them can work in combination.

1.1 Inadequate Experience and/or Initial Training

This factor was a major issue during the 1950–1980 period, when breccia styles had received very limited research description, and as a consequence were not included in many tertiary curricula. The knowledge problem was particularly acute in the United Kingdom/Europe, Australian, Canadian and Asian circuits. The problem still exists but has been alleviated by increasing published documentation.

1.2 The Clay/Mine Dump Problem

Many intrusive breccia columns are overprinted by later hydrothermal activity producing large amounts of clay or clay-associated minerals. Two basic mechanisms are operative, and in many cases both work in tandem to achieve a camouflage effect regarding surface exposures.

a. Late-stage overprinting by major hypogene clay alteration. This is achieved via alteration of both fragments and matrix to "clay minerals", and is particularly prevalent within deposit styles involving intermediate or advanced argillic alteration. The clay effect is even more enhanced where the rocks have been previously feldspathised, a common story in porphyry style mineralisation.
b. Late-stage overprinting via major supergene clay alteration. This is extremely common and occurs where normal weathering causes oxidation of sulphides (usually pyrite) and creates a large-scale, low-temperature "acid bath". The general effect is to convert silicates to kaolin and smectite style minerals. This effect is commonly superimposed upon the initial "hypogene" clay environment described above.

The net result is to produce rocks composed of soft white end-products, which within regions of moderate to high rainfall may become a clay-like mud at surface. Within early- or late-stage mining, the clay-like mud is easily confused with soil profile and/or old dumps/tailings. Although clay alteration acts to conceal intrusive breccias, it can also operate in reverse to the experienced breccia hunter, who should automatically go and check any clay-rich zone.

1.3 The Alteration Problem

Many intrusive breccias are overprinted by one or more hydrothermal stages which may cause pervasive alteration, such as feldspathisation, biotitisation, sericitisation, silicification or "skarn" development. This effectively conceals the original texture from casual observation. The problem is common, and any major mass of alteration should come under suspicion as being breccia based. In most cases slabbing with a rock saw may reveal some texture, and if possible a general investigation of more exposures or drill core will reveal areas that are less overprinted which reveal original rock texture.

1.4 The Volcaniclastic/Sediment Problem

Many intrusive breccias are missed in volcaniclastic terrains ,with both exposures and drill core materials being misidentified as agglomerates and tuffaceous rocks. This is one of the more difficult problems as many species of intrusive breccia are located within volcanic/subvolcanic terrains. Geologists are familiar with volcaniclastics and tend to record them automatically without considering that there may be intrusive breccias hiding within them. It can also

be very difficult to distinguish between the options without additional information such as the general stratigraphic column, layering and graded bedding. The situation is commonly further confused by surficial or hypogene alteration affects.

Similar problems can arise with sediments. The writer once mapped a circular patch of conglomerate within granite at the Ardlethan tin mine, and beyond thinking it mildly unusual, completely failed to recognise an intrusive breccia pipe. Initial identification of extensive fragmental rocks as a talus slope red-bed environment at the now well-established intrusive breccia pipe at Olympic Dam, South Australia, similarly caused serious geological misinterpretation. It is probably best practice in mineralised fragmental rocks to suspect any fragmental rock as a possible intrusive breccia. This approach at least safeguards against major errors, and stimulates the observer to gather more data.

1.5 Observational Problems

Breccias that are hidden beneath alteration overprints of all kinds commonly remain unrecognised despite the fact that they have quite characteristic knobbly textures. This broad-scale texture is usually best observed by standing well back from the questionable exposures (usually mineralised zones) and gaining a perspective of the rock types. Breccia zones usually stand out from the solid wall rocks –

1. With a change in texture from relatively solid blocky and/or layered rocks to knobbly zones associated with alteration/mineralisation
2. By a sudden loss of jointing, passing from normal country rock to "broken" breccia-zones.

The combination of the above two criteria can be utilised at all scales of observation, and it is commonly possible to spot or suspect the presence of major breccia from 0.5–1 km distance.

◻ Plate 1

Instant recognition
No recognition problem, classic format with mixed fragments and fragment rounding.
(Photograph provided by P. J. Pollard).
Wheal Remfry china-clay mine, Cornwall, England.

In anticipation of this volume being utilized by people who have little or no concept of the "intrusive breccia" style, this plate has been selected as a very visual, sizeable end-member of the clan. It is literally a black and white example with the bench face being some 5–6 m high. The body occurs within a white kaolin clay pit and measures some 400 m × 40–100 m, with around 90 m of vertical exposure. The contacts display many examples of breccia apophyses and veinlets, and onsite the fragments appear to have suffered both lateral and vertical transport. The plate illustrates very visible fragmented rock with clasts ranging from metre to low centimetre scales. The fragments are of different rock types with white kaolinised granite being more obvious than the less abundant dark metasediments.

Even at this distance many of the white fragments are distinctly ovoid to subrounded and the generally rubbly effect suggests a high proportion of cobble-like boulders. A milled-intrusive style breccia would be an immediate suspicion. The dark colour represents tourmaline alteration via a hydrothermal fluid which has been particularly effective at transforming small scale fragments and the grit/silt size matrix to tourmaline-silica altered rocks. (*Allman-Ward et al., 1982*).

Tourmaline-dominated intrusive style breccias are extremely common internationally, and are widely reported within porphyry-level domains of tin and copper systems (*Sillitoe, 1985*). They are commonly very early-stage features of magmatic-dominated systems and range up to sub-kilometre dimensions, commonly occurring in clusters. Tourmaline breccia systems are commonly disappointingly barren of metal content, but may provide a focus for later overprinting mineralising stages. They are also noteworthy in that a good case can usually be made for alteration being essentially synchronous with breccia formation implying actual breccia formation via a rapidly exsolving boron-rich fluid. In most non-tourmaline breccia of this style, the major alteration effects are clearly post breccia formation. Further Wheal Remfry examples are shown in ***Plates 12, 13, 29, 31–34.***

◻ Plate 2

Initial recognition
Minor problem of recognition due to fragment scale.
Kidston gold mine, Queensland, Australia.

This plate had been included as a second reasonably black and white example of an intrusive style breccia. The fragments are composed of dark coloured metamorphic rocks, along with less obvious pale coloured granite. Although recognition is reasonably easy, it is worth noting the general knobbly effect on the bench faces. This is very typical of intrusive breccia and a very useful long and short range criterion, where fragments are obscured by subsequent alteration or general dust cover from mining operations. The large central lensoid fragment of metamorphic host rock is particularly striking, and subrounded fragments of this scale are common within this major kilometre scale breccia pipe. Recognition of large-scale fragments is always difficult both at bench and drill core scales, and stand back observation is always recommended. At this mine a whole wall of the pit proved to be a fragment or screen, and substantive additional ore was ultimately located requiring expansion of the pit. The gold mineralisation at Kidston is largely contained within late-stage sheeted quartz veins ± pyrite, sphalerite, galena, which overprint both the breccia and later-stage "rhyolitic" dykes. The breccia pipe varies substantively in fragment size. The actual pipe is some 1.100 × 900 m at surface and extends subvertically for at least 300 m. At least two stages of intrusive brecciation are recognised, and there are indications of associated collapse breccia. The fluid history is dominated by magmatic contributions, although the exact nature of the fluids responsible for brecciation is unknown (*Baker and Andrews, 1991*). Further Kidston breccia textures are depicted in ***Plates 24, 25.***

PLATE 1 Instant recognition
No recognition problem, classic format with mixed fragments and fragment rounding.
Wheal Remfry china clay mine, Cornwall, England.
(Photograph provided by P. J. Pollard).

Initial recognition
Minor problem of recognition due to fragment scale.
PLATE 2 Kidston gold mine, Queensland, Australia.

Plate 3

Initial recognition
The clay/mine dump problem, breccia concealment and/or clues for recognition
Tin mine (non loc.), Bolivia.

This is a good example of breccia concealment, although in reality the situation would act as a focus for intrusive breccia hunters. The main clue here is the white zone (centre) of relatively fine-grained rubble, leading upwards into a white in-situ clay zone, forming part of the pit wall. Upon close inspection this was identified as a small well-milled intrusive breccia. The tendency to rill and create sandy/scaly scree zones is commonly a direct breccia signpost. (See *Plates 4, 6, 7*).

Plate 4

Initial recognition
The clay/mine dump problem – breccia concealment
Tin mine (non loc.), Bolivia.

A perfect example of intrusive breccia concealment which was overlooked by three experienced breccia geologists during some four hours of geological inspection. The sizable excavation consists of very soft white clay zones, (left) interspersed with zones of slightly more solid white kaolinised granitic rocks. (centre-left). The islands of rock contain quartz-cassiterite vein systems, whereas the excavated zones are milled kaolinised breccia containing particles of quartz-cassiterite. The soft material has been sluiced and is part of a large easily extractable breccia pipe system. Local information regarded the soft clay zones as waste dumps which had slumped into the pit!

Plate 5

Initial recognition
The clay problem, breccia concealment. Tin mine (non loc.), Bolivia.

A closer view of part of the pit wall from *Plate 4*. This again illustrates the difficulty of locating milled breccia within extremely weathered environments. The kaolinised milled breccia is located to the centre and right (whiter zone), with the broken fractured wall rocks occurring to the left (yellow jarositic staining). The lesson is to suspect all clay zones as potential intrusive breccia. The fragment size is not visible here, but the milled rock fragments rarely exceed 3–5 centimetres, and are contained within a matrix of silt/clay size materials.

PLATE 3 Initial recognition
The clay/ mine dump problem, breccia concealment and/or clues for recognition
Tin mine (non loc.), Bolivia.

PLATE 4 Initial recognition
The clay/ mine dump problem, breccia concealment
Tin mine (non loc.), Bolivia.

Initial recognition
The clay problem, breccia concealment
PLATE 5 Tin mine (non loc.), Bolivia.

◻ Plate 6

Initial recognition
The clay/breccia problem, breccia concealment and/or clues to breccia location.
Escondida copper mine, Chile.

The plate is a good illustration of how the tendency towards clay alteration within intrusive breccias can be utilized to locate potential candidates at long range. This picture was taken around 1997, and subsequent to local comments concerning bad ground, drilling and collapsing bench faces, an inspection confirmed kaolinised intrusive breccia in the right-hand back corner of the pit. (See *Plate 7*).

◻ Plate 7

Initial recognition
The clay problem, breccia concealment and/or clues to location.
Escondida copper mine, Chile.

This is a close-up picture of a bench face within the bottom right hand corner of the Escondida porphyry copper pit as shown in *Plate 6*. Features to note are the very rubbly, knobbly nature of the exposure, with an obvious tendency to fall down and disaggregate to form fine-textured scree (bottom of bench). The intrusive breccia had not been recognised at the time of photography (1997), and with further excavation expanded to a major breccia pipe body occupying the entire pit base. It is a late-stage intrusive breccia overprinted by advanced argillic alteration assemblages (alunite, pyrophyllite ± kaolinite, illite) and primary high-sulphidation copper mineralogy (chalcocite, enargite, bornite). Extensive pyrite-associated with the early porphyry mineralisation and the later high sulphidation system has oxidized to create acid groundwaters, resulting in supergene chalcocite enrichment and further supergene kaolin/smectite development in the altered porphyritic rock mass.

PLATE 6 Initial recognition
The clay/ breccia problem, breccia concealment and/or clues to breccia location.
Escondida copper mine, Chile

Initial recognition
The clay problem, breccia concealment and/or clues to location.
PLATE 7 Escondida copper mine, Chile

Plate 8

Initial recognition
Observational problems, gossanous exposures, textural changes.
Doi Ngom tungsten mine, Thailand.

This plate depicts a fairly typical long distance view of an intrusive breccia system, which at the time of photography had not been recognised due to a combination of gossanous iron-oxide overprinting and local inexperience. This is a common situation. The points to note here are

1 Always make sure to conduct initial observation from long range, and
2 Examine the excavation for both colour and textural changes. The latter should include noting knobbly-broken rocks and paying particular attention to the distribution of jointing. Brecciation destroys jointing and this feature commonly enables recognition from long range (100 m to 1 km).

In this example, large blocky country rocks with well developed through-going jointing (top right) transition into more featureless, less well-jointed materials (centre and left). The transition is also marked by a general colour change from a white-red-yellow combination (right) to a more uniform yellow less well-jointed zone, and breccia suspicions are immediately aroused. (See ***Plate 9***).

Plate 9

Initial recognition
Observational problems – knobbly-broken rocks and textural variations.
Doi Ngom tungsten mine, Thailand.

The yellow relatively featureless zone in the previous plate is shown here at close range, and can immediately be seen as extremely knobbly and without through-going joints. It is obviously a leached-gossanous exposure and equally obviously is the main focus of excavation.

Readers are strongly recommended to observe the textural changes within the exposure via a mental process of picking out zones that are more broken (more knobbly) versus zones which are less broken (more solid). The eyes will slowly attune, and eventually delineate large steeply inclined relatively unbroken slabs, surrounded by zones of extremely broken finer textured rubble. The big blocks are well seen to the top left and centre right, dipping some seventy degrees to the right. Although not easily visible, the rubble zones contain many rounded rocks arousing suspicions of intrusive style breccia. This intrusive breccia zone can be traced as an elongate body within metasedimentary host rocks, and several other mineral occurrences in the near vicinity suggest the possibility of a clustered breccia-pipe system relating to an intrusive. This is somewhat different to the original interpretation of oxidized vein systems.

PLATE 8 Initial recognition
Observational problems, gossanous exposures, textural changes
Doi Ngom tungsten mine, Thailand.

Initial recognition
Observational problems, knobbly-broken rocks and textural variations
PLATE 9 Doi Ngom tungsten mine, Thailand.

Plate 10

Initial recognition
textural concealment due to major overprinting alteration.
Ardlethan tin mine New South Wales, Australia. (Ardwest South Cherry pit).
(Photograph provided by D. W. Clarke).

The pit view illustrates an ore zone which had been under excavation for some 5–10 years before the intrusive-breccia character was recognised. In retrospect some major clues are present even at the broad-scale. The host rocks to the right are well-jointed relatively unaltered granite and the sun can be seen reflecting from the high angle flat joint surfaces of the pale red oxidised rocks.

Traversing to the centre and left the rocks become slightly redder within the ore zone, and the large joint faces vanish. The red colour is due to oxidation of sulphides and gangue minerals. The combination of joint loss plus major alteration is enough to raise suspicion of breccia, and further examination (***Plates 28ABC and 30***) revealed a major series of intrusive breccia pipes resulting in a total reorientation of exploration and mining development approaches.

Plate 11

Initial recognition
confusion with sediments
Olympic Dam (copper, gold, uranium) mine, South Australia, Australia
(Photograph provided by P. J. Pollard).

The now famous major intrusive breccia system at Olympic Dam provides another example where initial recognition of intrusive breccia proved very difficult. The plate illustrates a small area of breccia composed of fragmented granite.

Fragment size is variable, as is the degree of fragment rounding and alteration. The red hematite alteration affects all fragment sizes, but is principally focused within the small fragments and sand/silt size breccia matrix. Later-stage hematite veins (dull grey metallic) overprint the main breccia and are themselves offset by minor late-stage faulting.

The plate depicts an underground face where good exposures confirm the intrusive breccia style. However, the initial core logging favoured a talus slope (red bed) sediment style.

PLATE 10 Initial recognition – textural concealment due to major overprinting alteration. Ardlethan tin mine New South Wales, Australia. (Ardwest South Cherry pit) (Photograph provided by D.W. Clarke).

PLATE 11 Initial recognition – confusion with sediments. Olympic Dam copper-gold-uranium mine, South Australia, Australia (Photograph provided by P. J. Pollard).

2 General Features – Range in Styles and Scale of Intrusive Breccia

This section is designed to introduce the reader to the extreme range in style and scale involved in examples of intrusive breccia. The linking factors are the concepts of significant fragment transport with attendant mixing and milling. However, within this simple context the end results may range from vein to breccia pipe scales, and exhibit wide variations in the proportions of fragments (large bits) versus matrix (smaller bits). Similarly the proportions of rounded (milled) versus angular (broken) fragments is infinitely variable, and all of the above parameters may be linked with fragment size and vary from one portion of a system to another. The selected plates only show a small sample of the range, and a more complete cover is available by perusing the complete volume. For instance the vein styles are better illustrated within the epithermal and special feature sections.

Plate 12

General features – range in styles and scales. Medium scale breccia pipe. Intrusion, dilation and comminution of wall rock. Variably milled. Wheal Remfry Cornwall, England. (Photograph provided by P. J. Pollard).

This plate has been included to demonstrate a relatively early stage in the development of an intrusive breccia. It also illustrates some commonly observed textural features. The exposure is interpreted to represent the forceful intrusion of a sandy slurry into a fractured host rock. Within this context, the breccia can be viewed as two separate but related components. The larger (hammer scale) blocks of granite (white) give a general impression of having been dilated and mildly rotated. There is a general impression that they can almost be fitted back together, rather like a jostled jig-saw puzzle. These blocks are mostly angular and some appear to be forcefully levered away from their neighbouring smaller pieces. The remainder exhibit a size range extending down to cobble-sand scales (matrix) associated with a dark component. On close inspection this proves to be tourmaline ± quartz which is present as an alteration of the matrix. The pebble-scale fragments within the matrix exhibit considerable rounding and the matrix component is interpreted as a pressure-driven comminuting slurry. The slurry was associated with breaking and dilation of the host rock. Readers are referred to plates (*1, 13, 29, 31–34*) which exhibit other features of this well-exposed breccia.

Plate 13

General features – range in styles and scales. Medium scale breccia pipe – marginal sheeted fracture – well milled. Wheal Remfry china clay mine, Cornwall, England. (Photograph provided by P. J. Pollard).

This plate illustrates one contact zone of the tourmaline breccia pipe. The main body is of medium scale with a plan dimension of 400 × 40–100 m, and further examples from this well-exposed breccia are shown in plate (*1, 12, 29, 31–34*). This plate gives some concept of scale and the general style is somewhat different to the zone depicted in ***Plate 1***. The jig-saw, mildly jostled effect has been replaced by fragments which seem considerably more detached and jumbled. Although it is a little difficult to see, some well-rounded specimens are present towards the basal sector.

The marginal zones of intrusive breccia systems are extremely variable in texture, and range from intensely fractured, through sheet fractured to essentially unfractured. It is usually difficult to judge how much of the fracture effect is due to the original structure in the host rocks versus the effects of breccia emplacement.

In this example, the narrow 1 m scale border zone contains a concentration of brittle sheeted fracture. The fractures contain veins (tourmaline?) and there is a relatively sharp contact with the jumbled breccia zone.

PLATE 12 General features – range in styles and scales. Medium scale breccia pipe. Intrusion, dilation and comminution of wall rock. Variably milled. Wheal Remfry Cornwall, England.
(Photograph provided by P. J. Pollard).

PLATE 13 General features – range in styles and scales. Medium scale breccia pipe – marginal sheeted fracture – well milled. Wheal Remfry china clay mine, Cornwall, England
(Photograph provided by P. J. Pollard).

◻ Plate 14

General features – range in styles and scale – mixed fragments, vague alignment of larger clasts, substantive matrix proportions and rounding of fragments – crackled wall rocks with evidence of forceful matrix intrusion. Medium- to large-scale breccia pipe. Dizon copper mine, Puo Diatreme, Philippines. (Photograph provided by G. W. Morrison).

This specimen displays the contact zone between a section of a large 400–500 metre diameter intrusive breccia (Puo Diatreme), and porphyritic andesite. The fragments (0.5–2 cm scales) exhibit a vague alignment and are composed of variably altered andesite(s) including grey – silicified, paler grey sericitized and white sericite/illite/kaolinite? Several fragments contain spots of pyrite and there are a few unidentified darker fragments. The fragment variability in contrast to the wall rock is suggestive of fragment transport and mixing, and along with the general alignment suggest an intrusive breccia interpretation. Further suspicions are aroused by the high proportion of fragment rounding, especially at the medium to lower size ranges, and the high degree of matrix components (40–60 %). The matrix has become mildly altered and both the matrix and smaller fragments are partially altered to dark sericite. Readers are reminded that alteration via entry of fluid through an unconsolidated matrix is very common in milled breccia, and results in blurring and ultimately replacement of the smaller particles before enchroaching upon the larger sizes. In this instance, matrix alteration is relatively mild, and only a few fragments or parts of fragments have been affected by the dark sericitisation. It is not possible to discern whether the dark sericitisation is produced by a fluid accompanying or postdating brecciation, although the latter is more common and further search may find zones of unaltered matrix.

The photograph also emphasizes the importance of noting the composition and alteration characteristics of intrusive breccia fragments. It is clear the pale sericitisation, silicification, sulphide alteration (pyrite) and the white (illite/kaolinite?) alteration precede both the breccia and dark sericite formation suggesting the nearby presence of a major evolving hydrothermal system.

The andesitic wall rocks are shattered, with an oriented brittle crackle pattern. The crackled zone is intruded by the matrix, which has levered off wall rocks into the main flow zone. The origin of the brittle fracturing is a little difficult to interpret with both pre-existing fracture and/or fracture caused by the intruding breccia being viable explanations.

The boundaries of intrusive breccia bodies vary considerably and range from sharp with little sign of wall rock fracture, to various styles of brittle fracture stockworking – sheeting and/or shearing. In the case of major-scale pipes it is always difficult to discern just how much of the edge structure is pre-existing or formed via the violent intrusive action of the breccia. In reality there is very little scientific structural documentation or analysis of this problem.

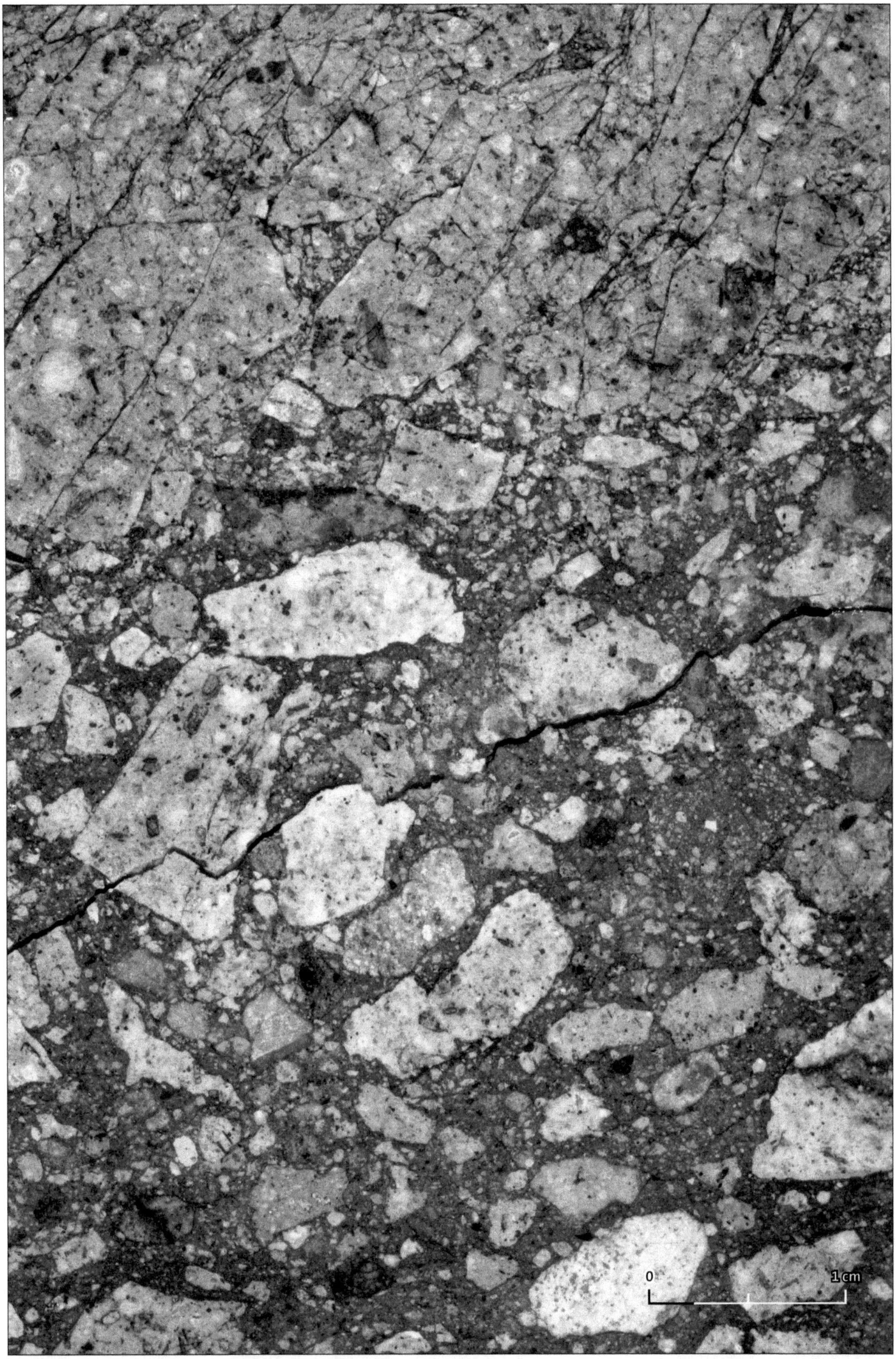

PLATE 14 General features – range in styles and scale – mixed fragments – vague alignment of larger clasts, substantive matrix proportions and rounding of fragments, crackled wall rocks with evidence of forceful matrix intrusion. Medium- to large-scale breccia pipe.
Dizon copper mine, Puo Diatreme, Philippines. (Photograph provided by G. W. Morrison).

Plate 15

**General features – range in styles and scale. Major-scale breccia pipe with highly comminuted style. Fragment mixing and abundant matrix component.
The El Teniente copper mine, Chile.
(Photograph provided by D. Andrews).**

The spectacular and large Braden breccia-pipe complex at the El Teniente mine has a surface diameter of some 1300 m and extends downwards in a tapering format in excess of 1600 m. It is interpreted to represent an extremely milled end-member of the intrusive breccia family, and in the plate opposite the entire rock mass is considered to have been in vigorous motion, such that even the 20–40 cm blocks of porphyritic igneous rocks (buff colour) are smoothly rounded. The darker fragments are andesite and some of the clasts contain pre-breccia mineralisation. Breccias of this style with abundant finer-grained components are commonly referred to as matrix-supported, although the amount of matrix may vary dramatically from place to place.

The enormous breccia complex is considered to have developed over a time span of greater than 2 million years, between >6.4 and 4.4 Ma *(Skewes et al., 2002).*

Plate 16

**General features – range in styles and scale. Major-scale breccia pipe with highly comminuted stylefragment mixing, abundant matrix component.
El Teniente copper mine, Chile.
(Photograph provided by D. Andrews).**

This plate illustrates a contrasting style and scale of milled brecciation in comparison to that exhibited in *Plate 15.*

It represents another portion of the Braden breccia system which lacks large fragments. Fragment size ranges from centimetre to millimetre scales, with abundant evidence of rounding, mixing and milling. It is worth noting that not all fragments are rounded. Angular fragments are common in intrusive breccias, and can be explained as:-

1. representing "recently" acquired wall rocks which have not had sufficient time to be abraded
2. fragments which were once larger and have only "recently" broken apart
3. a fragmentation style relating to explosive forces and rapid injection of debris.

The matrix (smaller scale fragments) is extremely visual down to grit size, at which point the particles become slightly vaguer and appear to be altered (dark tourmaline?). The effects of hydrothermal fluids are usually first manifested by alteration of the finest breccia (matrix) particles and these are the focal observation point for the observer. Larger fragments are commonly unaffected in tourmaline breccias.

PLATE 15 General features – range in styles and scale,. Major scale breccia pipe with highly comminuted style, fragment mixing and abundant matrix component. The El Teniente copper mine, Chile (Photograph provided by D. Andrews).

PLATE 16 General features – range in styles and scale. Major scale breccia pipe with highly comminuted style, fragment mixing, abundant matrix component. El Teniente copper mine, Chile. (Photograph provided by D. Andrews).

Plate 17

General features – range in styles and scale. Small-scale pebble dykes. Tintic district, Utah, United States of America. (Photograph provided by D. Andrews).

Intrusive style brecciation adopts a wide variety of formats and is by no means restricted to large-scale circular pipes. The general process also produces lensoid pipes, dykes of various scales, and even milled segments within veins. The latter are common in epithermal domains. A species which always attracts attention and is generally very visible, takes the form of narrow (0.5–1 m scale) dykes with very rounded/milled fragments at pebble sizes. Not surprisingly these are universally called pebble dykes. There is occasional debate as to the amount of fragment movement as opposed to "spinning on the spot", but in general some upward movement can be demonstrated. The example shown here is 0.5 m scale.

Plate 18

General features of intrusive breccia – range in size and scale. Small-scale pebble dykes. Tintic district, Utah, United States of America. (Photograph provided by D. Andrews).

This close up photograph of the pebble dyke shown within ***Plate 17***, illustrates how the name pebble dyke has arisen. The milled fragments are set within a finer-grained, largely unconsolidated matrix, and hence tend to weather out. The fragments are extremely rounded. The larger clasts are quartz-rich, and although not visible here the fragments are composed of quartz veins, quartzite, shale, limestone, and pyrite.

Pebble-dykes are commonly late-stage features within porphyry systems, and not overprinted by multiple hydrothermal events. They are usually post-mineralisation events, although this could be a little illusory, as they are more difficult to discern when mineralised and/or altered. Models involving escaping/expanding hydrothermal fluids (gas and/or liquid) channelled through existing structures are generally proposed to explain their characteristics. (e.g. *Bryant, 1968*).

PLATE 17 General features – range in styles and scale.
Small-scale pebble dykes
Tintic district, Utah, United States of America.
(Photograph provided by D. Andrews).

General features of intrusive breccia – range in size and scale. Small-scale pebble dykes.
Tintic district, Utah, United States of America
PLATE 18 (Photograph provided by D. Andrews).

◻ Plate 19

General features – ranges in styles and scale. Medium-scale breccia pipe, with small-scale fragments, polished quartz pellets.
Llallagua tin mine, Bolivia.
(Photograph provided by P. J. Pollard).

This plate illustrates a common situation with intrusive breccias, where intense milling has created small scale fragments with considerable fine-grained matrix. In this instance the breccia exposure is substantive (20–30 m), and the specimen consists of relatively few clay-altered (white) igneous fragments, and a large matrix component composed principally of siliceous sediments (glassy grey). The well-rounded fragments have polished surfaces assumed to be related to comminution. The breccia is unconsolidated, post-mineralisation and within the high rainfall environment is both waterlogged and soft. It resembles a mine tailings dump and this style is commonly overlooked and misclassified as surficial weathering debris. (See also *Plates 3–7*).

◻ Plate 20

General features – ranges in styles and scale. Small-scale breccia dyke with small scale fragments, polished.
Boska tin mine, Herberton, Queensland, Australia.

Strongly milled breccia with fragment size at the 0.5–1.5 cm scale. The fragments are extremely well-rounded (pellets), highly polished quartz grains probably representing original quartz phenocrysts. They are hematite coated and stained due to surficial iron-oxide derived from weathering of the biotite-altered matrix. This breccia is very irregular and of very small-scale (2–20 cm), however it remained unrecognised over some 50 years of local mining, and when finally identified proved to be one of several in the adjacent tin field, with major implications for breccia-pipe exploration.

PLATE 19 General features – ranges in styles and scale. Medium-scale breccia pipe, with small-scale fragments, polished quartz pellets
Llallagua tin mine, Bolivia.
(Photograph provided by P. J. Pollard).

General features – ranges in styles and scale. Small-scale breccia dyke with small-scale fragments, polished quartz pellets
PLATE 20 Boska tin mine, Herberton, Queensland, Australia.

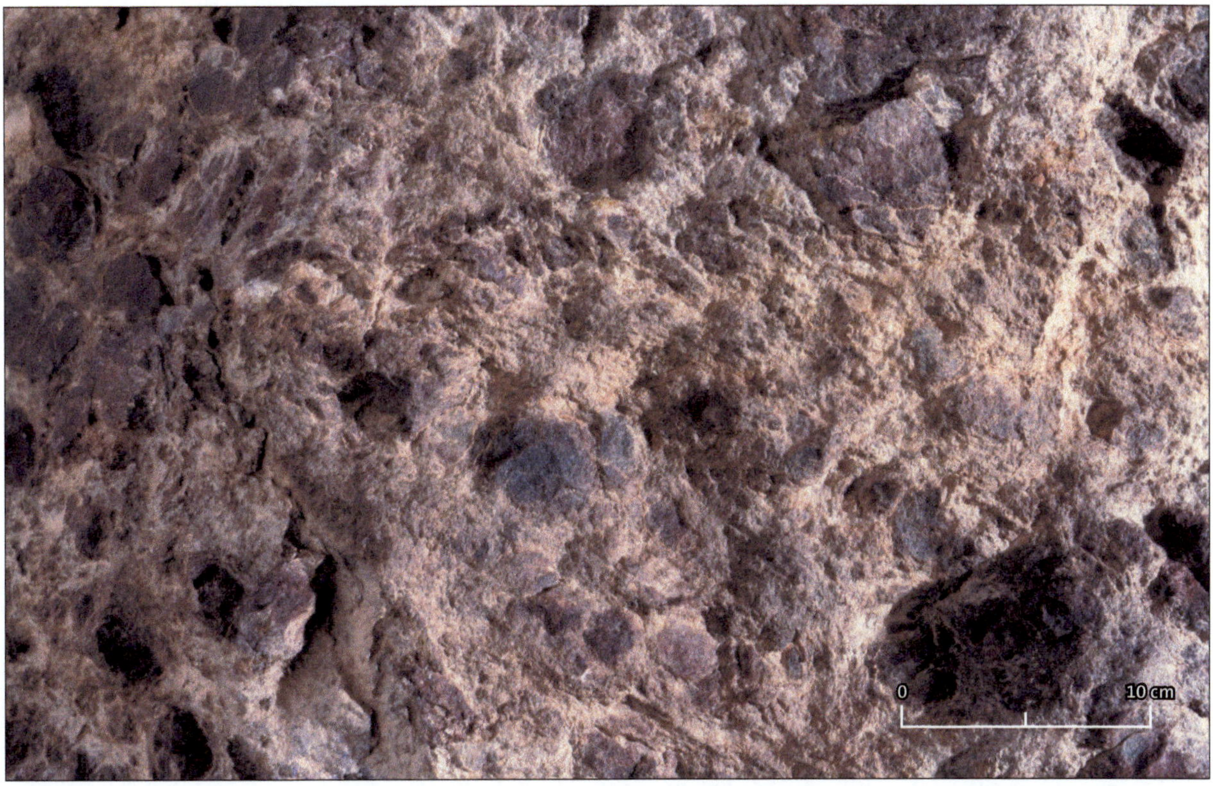

Plate 21

General features of intrusive brecciation – ranges in styles and scale. Tuffisite dyke. Mt Leyshon gold mine, Queensland, Australia.

Intrusive style breccias occur at a wide range of scales, and this figure illustrates an extremely fine -scale example.

A The central unspectacular white zone represents a high angle dyke, which traverses some 50 m of pit walls at Mt Leyshon. The wall rocks are intrusive breccia which form part of a major kilometre-scale pipe-like system. The knobbly nature of the wall rock is just discernible. Mineralisation is a late-stage event overprinting all the rock types. The white material is soft, clay-like and referred to locally as tuffisite. The "clay" alteration is probably a primary rather than a supergene effect.

B The white-clay dyke depicted within ***Plate 21A***, is not uniformly textured, and contains discrete zones of visibly fragmental materials (base) and finer-grained zones (top). The coarser materials are composed of multiple fragment types at the 0.5 cm scale ranging down to submillimetre "dust" particles. Some of the larger fragments have clearly been affected by hydrothermal alteration prior to inclusion within the rock (dark-sericite/biotite?, white clay?). The fragments are predominately rounded, and if this plate is compared with many others within this book, it will be immediately obvious that the only real textural difference is one of scale. The grain size of this rock attracts the name "tuffisite" rather than "pebble dyke" or milled breccia. Without the field context a standard volcaniclastic name would be appropriate, again illustrating identification problems at drill core or restricted outcrop scales. The "tuffisite" could be viewed as a fine-grained milled matrix component which has become separated ("elutriated") from a more substantive breccia at depth. This concept is illustrated in ***Plate 35***.

C The fine-grained rock depicted within ***Plate 21B***, as a relatively uniform white zone (upper region) is actually clearly fragmental when examined in thin section. The problems of scale now become very obvious, as the texture of multiple fragment types, fragment rounding with significant matrix component is typical of intrusive breccia. It is equally clear that without good field context, any intrusive breccia component is easily overlooked in favour of volcanic-volcaniclastic interpretations. It is not surprising that many intrusive breccia systems are initially overlooked when coping with volcanic wall rocks.

PLATE 21 A

PLATE 21 General features of intrusive brecciation – ranges in styles and scale – tuffisite dyke. Mt Leyshon gold mine, Queensland, Australia.

PLATE 21 B

PLATE 21 C

3 Multistage Structural and Hydrothermal Overprinting

Multistage intrusive breccias are components of evolving hydrothermal systems, with a tendency to form either early or late within the order of events. The early-formed styles are subject to refracturing and introduction of new hydrothermal fluids, and it is also not unusual to find more than one stage of intrusive breccia formation.

The pattern of nearly all hydrothermal deposits involves a succession of fracturing events coordinated in time with a succession of hydrothermal fluids (paragenetic development). It follows that hydrothermal intrusive breccias are commonly overprinted by one or more later stages. In many cases the overprinting and associated alteration may become so complex and/or intense that the original breccia texture is difficult to discern. This section introduces the reader to some of the textural problems created, and stresses the need for very careful observation. Once again there are many other examples scattered throughout this volume and this section is only intended as a focus for overprinting issues.

◻ Plate 22

Multistage structural and hydrothermal overprinting.
Cuajone (porphyry copper) mine, Peru.
(Specimen provided by D. Andrews).

This specimen illustrates a very common situation in intrusive breccia domains, where the main breccia contains fragments which have been previously brecciated. The early brecciation is best displayed in the large fragment (left) and an early intrusive breccia event is suggested with some evidence for fragment rounding and a major matrix component. The latter is clearly of fine grain size, although details have been obscured by the silica-sericite (grey) alteration effects.

The main breccia is another example of an intrusive breccia style which is common in tourmaline-related environments. This style contains a disproportionate amount of very angular, splinter-like fragments. The matrix commonly exhibits a fine dust/sand-scale matrix substantively obscured by tourmaline alteration.

The lack of mineral fragment rounding is a little difficult to explain. One possibility is that the forceful intrusive activity is at the explosive end of an "expanding gas" scale. This concept envisages a short sharp forceful intrusion/fragmentation of matrix materials with limited time to achieve substantive milling. The multistage development is completed by the pyrite (yellow) which although not totally clear in parts of the plate, is infill occupying a late-stage brittle fracturing of the main breccia (see fracture filling at base).

◻ Plate 23

Multiple structural and hydrothermal overprinting.
Esis (porphyry copper) prospect, East New Britain, Papua New Guinea.

This deceptive intrusive breccia is a very good illustration to remind the observer to carefully assess the fragments for signs of previous brecciation/mineralisation. The most critical observation within this rock is the fragment (lower centre-right) of porphyritic rock with a dark vein traversing vertically (left hand side). This is composed of biotite ± pyrite, and would classify as potassic "alteration" indicating the presence of a nearby porphyry copper system. A signal of this magnitude cannot afford to be overlooked. The main breccia resembles a volcaniclastic rock and this sample falls into the "recognition problem" category. In the field its pipe-like context is apparent, however as an isolated specimen it can only be placed under suspicion due to:

1. A wide range of fragment compositions
2. Extreme fragment rounding, which has affected most of the larger pieces
3. A high matrix component, with little or no sign of layering and/or graded bedding.

The matrix is very mildly altered and the pale green, pale yellow colours are due to chlorite and epidote, representing a weakly developed propylitic alteration.

PLATE 22 Multistage structural and hydrothermal overprinting.
Cuajone (porphyry copper) mine, Peru.
(Specimen provided by D. Andrews).

Multiple structural and hydrothermal overprinting.
PLATE 23 Esis (porphyry copper) prospect, East New Britain, Papua New Guinea.

Plate 24

Multistage structural and hydrothermal overprinting.
Kidston gold mine, Queensland, Australia.

The specimen illustrates a common theme of intrusive brecciation where several generations of hydrothermal activity are finally overprinted by brittle fracture-controlled hydrothermal fluid flow accompanied by precipitation of economically valuable mineralogy. In this case the bulk of the Kidston gold credits are contained in peripheral sheeted quartz vein systems. The quartz (white centre) is accompanied by sporadic sphalerite (dark), pyrite (dull yellow) and gold (not visible). The early stage intrusive breccia contains a mixture of fragments including pale fine-grained porphyritic rocks (left), very altered originally granitic rocks (centre-base) and quartz vein fragments (top-centre).

The original nature of the finer-grained matrix has been obscured by several stages of overprinting alteration, and fragment matrix relationships are obscure. Utilising data available from other samples the sequence is interpreted to contain early stage pre-brecciation K-feldspar alteration (see *Plate 25*), which is suspected here in some of the paler granitic zones. This has been extensively overprinted subsequent to brecciation by the dark silica ± sericite alteration of the gritty matrix (basal region). The relationship of the coarse-grained pyrite (centre right) to the quartz vein is obscure, but it does appear to overprint the dark sericite. Complex relationships of this nature are typical and multiple samples are required to resolve their timing.

A most important texture within a porphyritic fragment (top right) is worth noting. The quartz "phenocrysts" exhibit a tendency to link together in a chain like manner, with the crystals all pointing in one direction. This is most probably a unidirectional solidification texture (UST), which forms in apical regions of many intrusives associated with mineralisation. The crystals project downwards into the underlying chamber. (See also *Plates 2, 25*).

Plate 25

Multistage structural and hydrothermal overprinting.
Kidston gold mine, Queensland, Australia.

The specimen illustrates a small portion of drill core from the kilometre diameter intrusive breccia complex (see also *Plates 2, 24*).

This sector of the pipe is composed of fragments at the 2–4 cm scale which are extremely tightly packed such that they are either touching one another, or just separated by smaller fragments/matrix at the 0.5–1 cm scale. There is very little fine-grained matrix. Terminology such as fragment-supported might be utilised in this situation, although the reader should be aware that this is a relative term and that textural variation from matrix to fragment supported is common within metre length scales in intrusive breccia.

The fragments exhibit minor rounding and are composed of gneissic rocks (dark, left), granitic rocks (pink and grey, base), very fine-grained (pale) quartz porphyry (centre right), and broken (mid grey) quartz veins (top left). The pale quartz porphyry has been overprinted by pink potassic alteration prior to brecciation and the dark metamorphic gneiss contains early quartz veins. A green tinge at various points represents late-stage epidote/chlorite alteration. In larger specimens this overprints the milled gritty matrix.

Two late stage quartz veins clearly post-date all previous events.

PLATE 24 Multistage structural and hydrothermal overprinting
Kidston gold mine, Queensland, Australia.

Multistage structural and hydrothermal overprinting
PLATE 25 Kidston gold mine, Queensland, Australia.

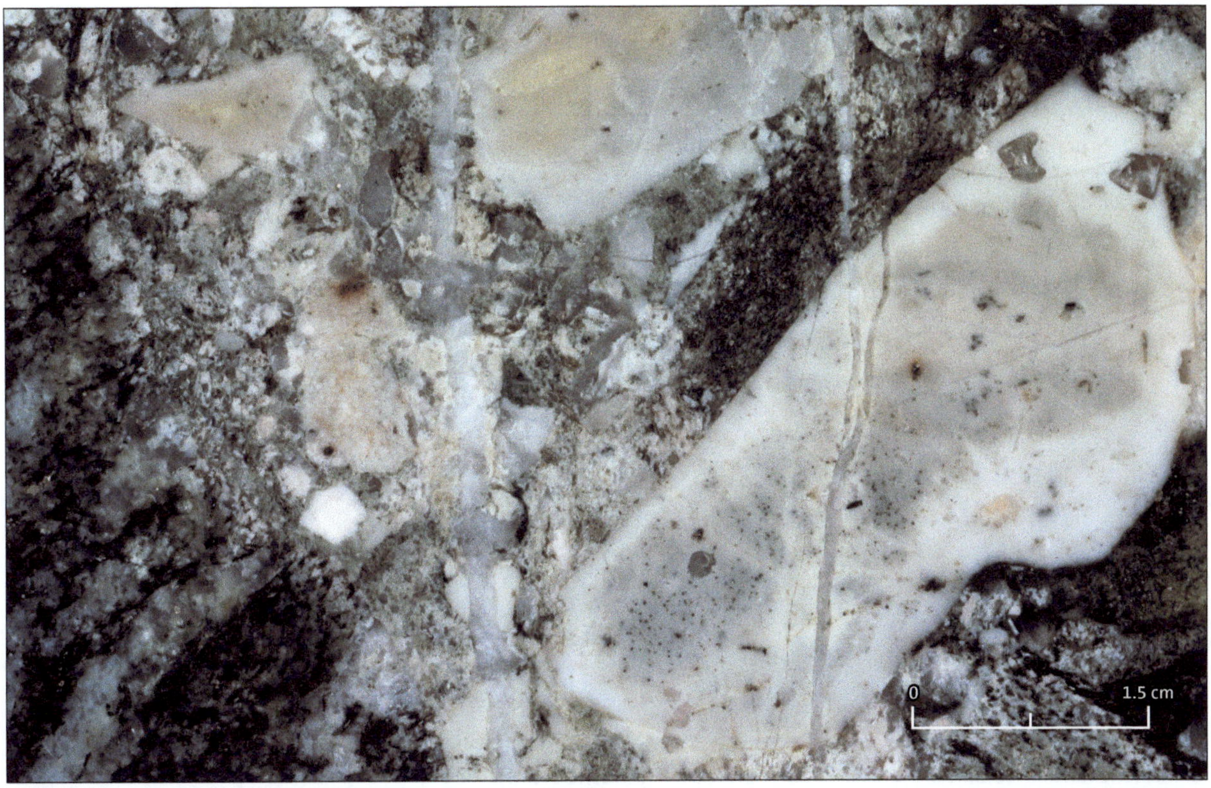

Plate 26

Multistage structural and hydrothermal overprinting – textural modification via alteration. Dalam fragmental unit Grasberg (copper, gold) mine, Papua, Indonesia.

The problem of multiple hydrothermal overprinting is extremely common within intrusive breccia systems, and original textures can be completely or partially obliterated. This becomes a major problem if geologists are trying to establish the original breccia details and also presents difficulties in unreveling paragenetic sequences.

The plate presented here is an excellent example of extreme complexity from the huge Grasberg system, where a kilometre scale breccia pipe has been so heavily modified by alteration that it is very difficult to establish its original parameters.

The fragmental nature is easily perceived, although all of the visible fragments have been modified by intense potassium feldspar (potassic) alteration (pale pink-white). It can just be discerned that they were once fine-grained porphyritic igneous rocks. Similarly, a reasonable degree of rounding could be interpreted. The nature of the matrix is impossible to assess, as it has been successively changed by

1. Intense potassium feldspar alteration. (The pale pink mottling between fragments to the top left, and bottom right).
2. Intense brown biotite alteration (the darker colours between fragments at bottom centre to bottom right)
3. Silicification ± sericite (pale grey and white) in a 0.5 cm zone either side of a central narrow pyrite vein (vertical vein to the left of centre).

The pyrite vein is cut by two small low angle anhydrite veins.

Plate 27

Multistage structural and hydrothermal overprinting. Ok Tedi (porphyry copper) mine, Papua New Guinea.

This is an excellent example of an intrusive breccia that proved extremely difficult to recognise due to overprinting alteration effects. The specimen has been affected by major potassium feldspar alteration. The alteration (pale pink-grey-white) is particularly effective at modifying and obscuring the details of the finer-scale matrix materials, and even the larger fragments have somewhat blurred outlines.

The original composition of the fragments is a little obscure but it is clear that several different lithologies are present, and that fragment rounding is well-developed.

Although the breccia texture is reasonably obvious here, it must be pointed out that the rock has been both slabbed and ground in an effort to bring out a suspected breccia texture. The original rock is a featureless white-grey rock with a proportion of the pale potassic alteration being converted to white supergene clay minerals. There are no obvious fragments, and breccia was only suspected by noting a mildly knobby surface within an obviously very altered rock. The darker lines are original sulphide-dominated veins now converted to chalcocite in the zone of supergene enrichment.

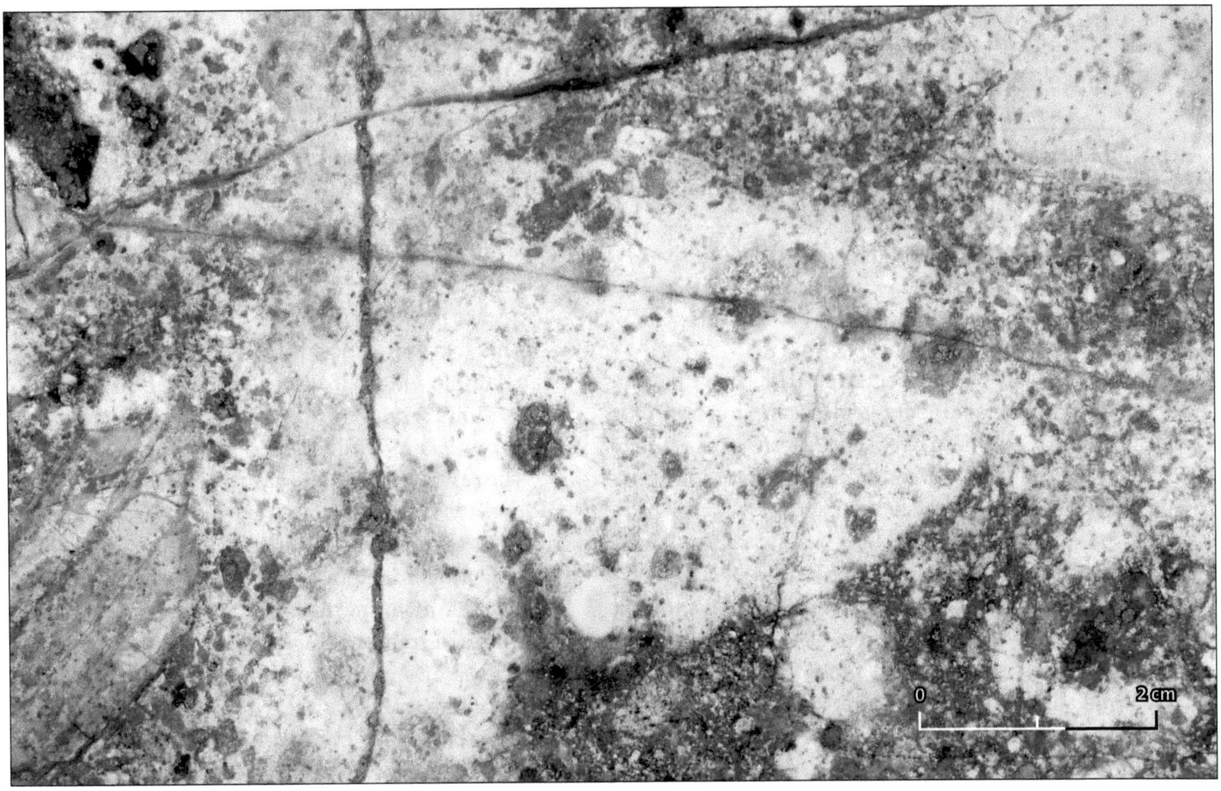

PLATE 26 Multistage structural and hydrothermal overprinting – textural modification via alteration. Dalam fragmental unit Grasberg (copper, gold) mine, Papua, Indonesia.

PLATE 27 Multistage structural and hydrothermal overprinting. Ok Tedi (porphyry copper) mine, Papua New Guinea.

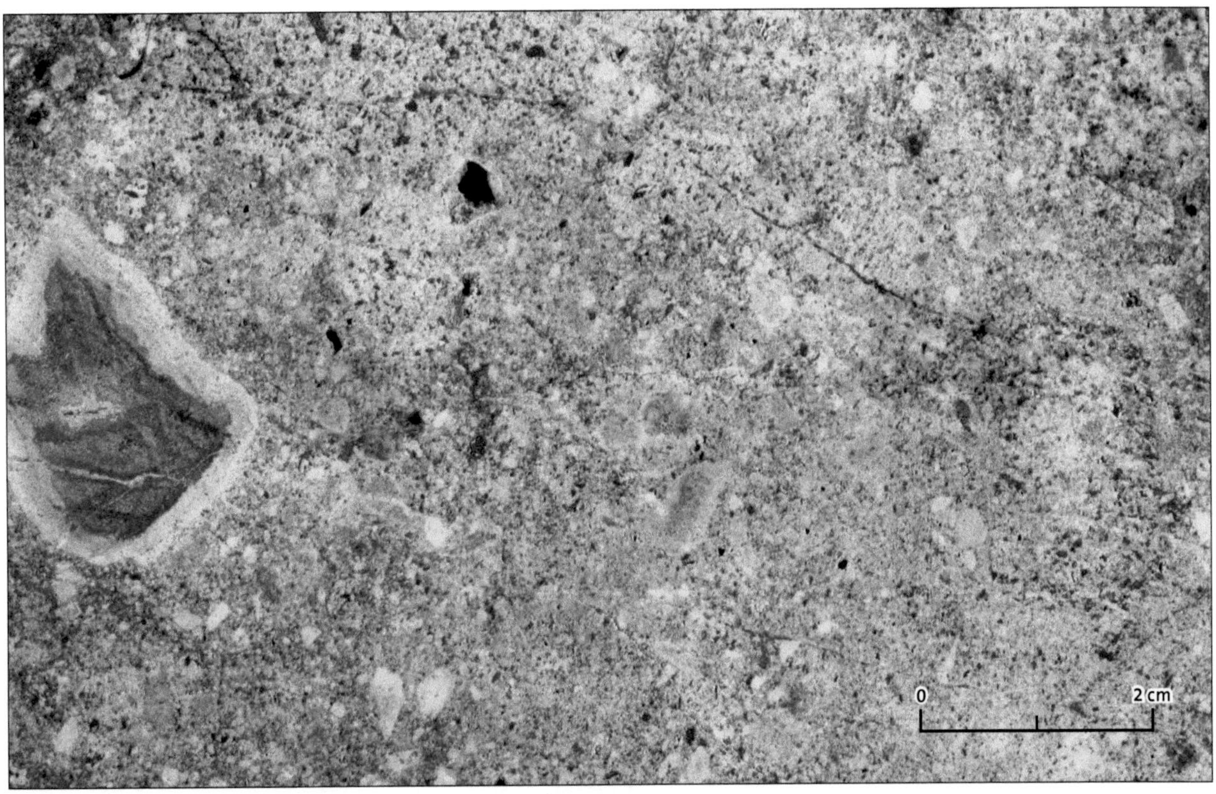

Plate 28 A B C

Multistage structural and hydrothermal overprinting – textural modification via intense alteration.
Ardlethan tine mine, New South Wales, Australia. (Ardwest – South Cherry Pit).
(Specimens provided by G. W. Clarke).

The intrusive breccia system at Ardlethan tin mine remained concealed from geological perception for many years. The main deposit (***Plate 10***) was vaguely designated disseminated and local exploration proceeded in an empirical fashion.

The main breccia pipes are heavily camouflaged by alteration. Some concept of the problem can be gained from inspecting ***Plate 28A***. This slab has been specially prepared to enable good vision, but it is seriously doubted that most first time viewers will be able to see any texture remotely resembling breccia. In reality the actual field exposures are even more obscure.

The problem posed by ***Plate 28A*** is relatively common in situations where granite is the only major host rock. The milling process commonly liberates the quartz component at approximately the same grain size as within the original rock. Consequently a milled mixture of quartz grains and powdered feldspars can look fairly similar to the original host. Any subsequent alteration preferentially affects the feldspar dust, again creating an effect resembling altered granite.

This has occurred within ***Plate 28A***, where the pale brown sericitic alteration is actually concentrated within milled matrix materials. The darker components are mostly tourmaline.

This textural concealment is revealed by etching the sample via hydrofluoric acid fumes (*NB. This process is very dangerous and should only be attempted under proper laboratory supervision*).

The result (***Plate 28B***), clearly reveals two separate domains. The fragments retain a coarser-grained igneous texture with the original feldspars standing out as white linked interstitial networks. The milled matrix is much finer-grained, and the combination of rounded fragments within a milled dusty/sandy matrix is enough to rouse suspicion of intrusive breccia. The smaller inset within ***Plate 28C***, reveals that fluid ingress is not restricted to the permeability of the matrix but is also via extremely small-scale discontinuous cracks. These are picked out as minute tourmaline rich infill zones running vertically up the centre of the inset. The brown colouration is due to surficial weathering and precipitation of iron oxides along a crack.

PLATE 28 C

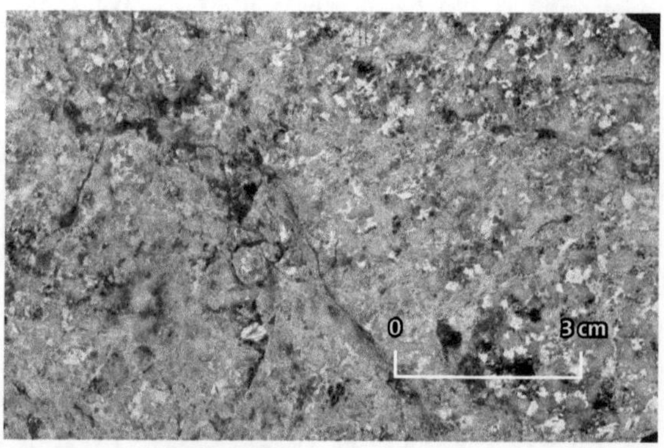

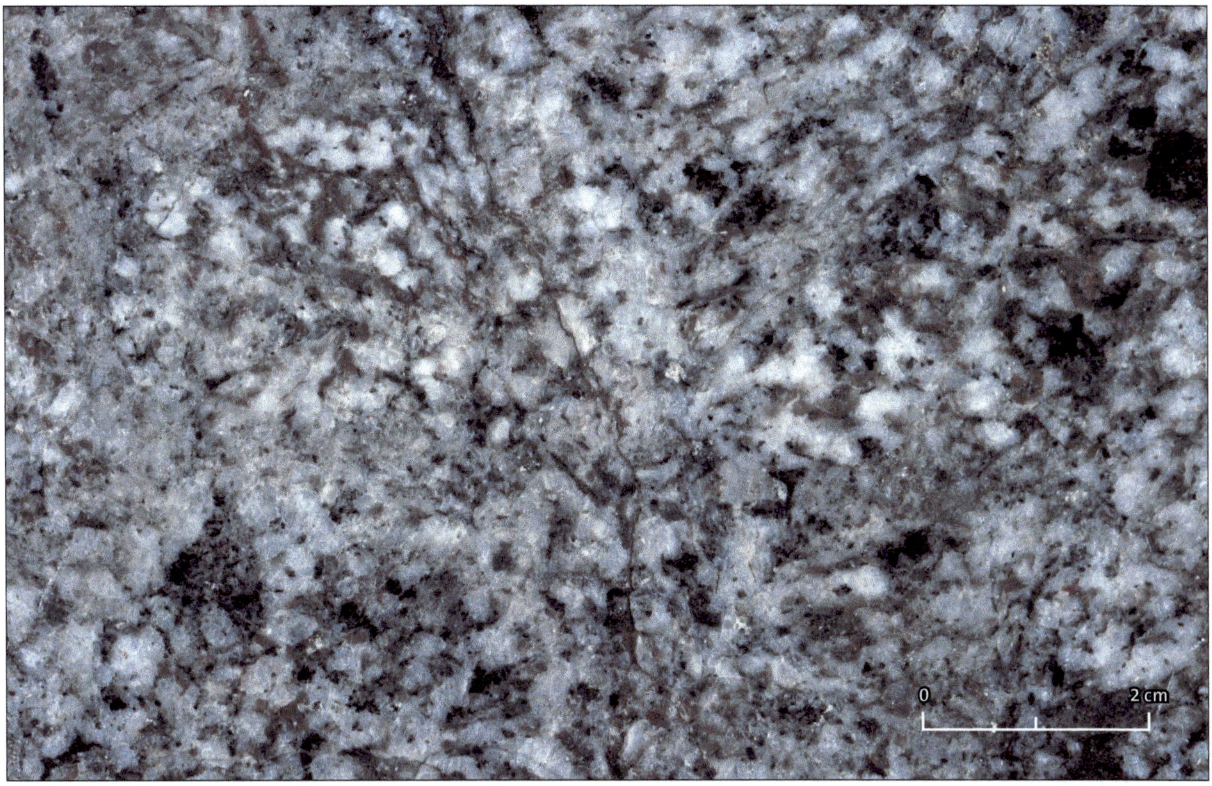

PLATE 28 A

PLATE 28 A B C
Multistage structural and hydrothermal overprinting – textural modification via intense alteration
Ardlethan tine mine, New South Wales, Australia. (Ardwest – South Cherry Pit)
(Specimens provided by G. W. Clarke).

PLATE 28 B

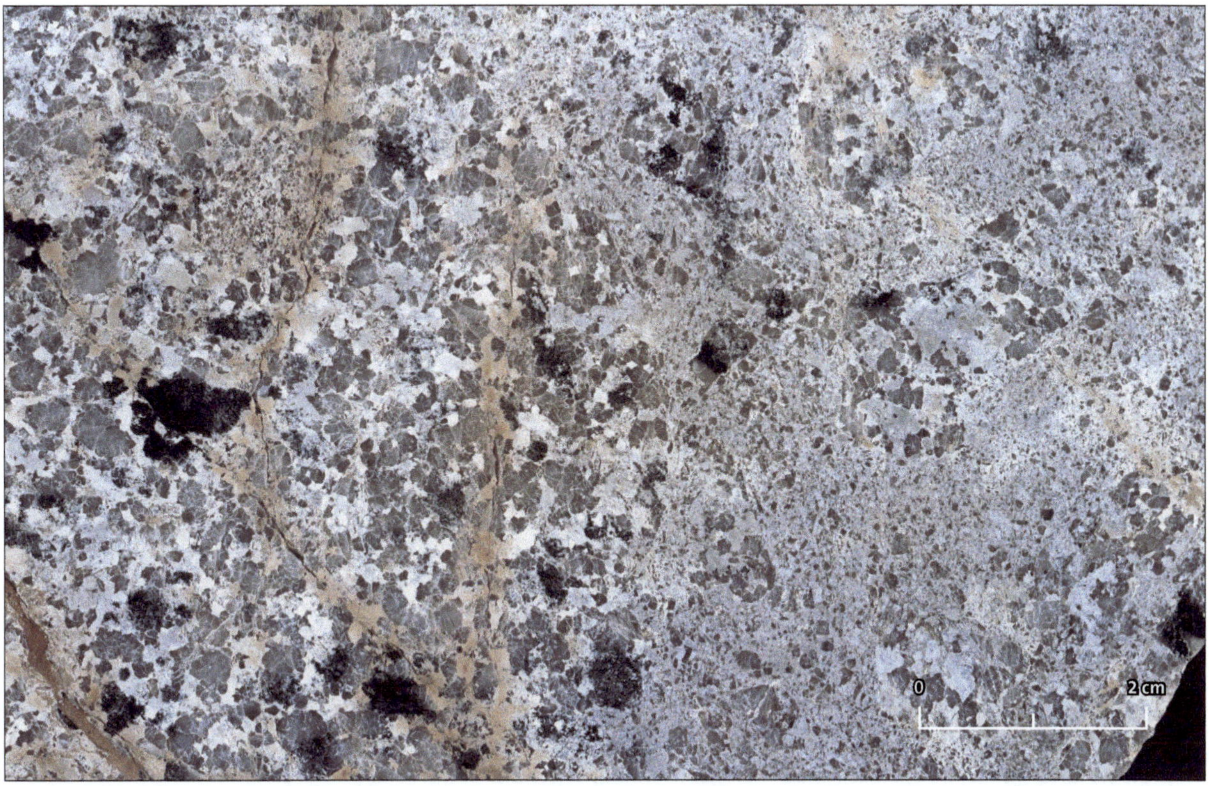

4 Special Features Associated with Selected Intrusive Breccias

This section is designed to illustrate some selected special textures which may be encountered within individual intrusive breccia systems.

Plate 29

Special features associated with selected intrusive breccias – intrusive breccia as veins.
Wheal Remfry china-clay mine, Cornwall, England.
(Specimen provided by D. Kirwin).

This plate has been included as a cautionary note to all field geologists. At first sight it would appear to represent a small-scale brittle fracture vein system disrupting a porphyritic igneous rock. However, a closer inspection (bottom centre) reveals that the subcentimetre scale veins are actually composed of finely milled rock dust, which is partially replaced by tourmaline (black). This is interpreted to suggest that the fracture system has been exploited and dilated by an intruding pressurized slurry, and the "veins" are conceptually more akin to dykes. This is probably an early stage of the process illustrated in *Plates 12 and 31.*

Situations of this nature are common in intrusive breccia systems, and are particularly important to recognise. An isolated exposure of this breed may be signalling the presence of a nearby or underlying major breccia pipe. The texture within the vein is a little difficult to see at this scale, but can be easily visualized via inspection of the altered matrix zones in P*lates 31–34*.

Plate 30

Special features associated with selected intrusive breccias – onion skin textures.
Ardlethan tin mine, New South Wales, Australia.
(Specimen provided by G. W. Clarke).

Geologists take great delight in using colloquial names to describe textures and these are usually only of value at the local scale.

This slide depicts some vague curved zones within a fragment. These are relatively common within granitic fragments and commonly attract the description of onion skin textures. The grey curved domains (centre and right) are sericitic alteration of the granite (pale). Presumably a minute curving discontinuous fracture had allowed entry of a hydrothermal fluid.

The main breccia zone (left), consists of a milled matrix of granitic composition which has been partially altered to tourmaline (dark).

PLATE 29 Special features associated with selected intrusive breccias – intrusive breccia as veins. Wheal Remfry china-clay mine, Cornwall, England (Specimen provided by D. Kirwin).

PLATE 30 Special features associated with selected intrusive breccias – onion skin textures. Ardlethan tin mine, New South Wales, Australia (Specimen provided by Dr. G.W. Clarke).

Plate 31

Special features associated with selected intrusive breccias – curved sheeted fractures and alleged decompressive shock textures.
Wheal Remfry china-clay mine. Cornwall, England.
(Photograph provided by P. J. Pollard).

The texture illustrated within this exposure is interpreted as representing a curved sheeted fracture system, being exploited and dilated by the forceful intrusion of pressurized milled intrusive breccia. The milled matrix has been converted to tourmaline (dark), and the situation is considered to result from the explosive escape of a boron-rich volatile fluid from an apical zone of a granite magma.

The origin of the curved fractures remains contentious, although decompressive shock exfoliation related to rapid pressure release is a widely accepted hypothesis *(Allman-Ward et al., 1982).* The rectangular network grid pattern present within ***Plate 29***, is also evident within parts of the larger curved system.

Plate 32

Special features associated with selected intrusive breccias – curved sheeted fractures and alleged decompressive shock textures.
Wheal Remfry china clay mine, Cornwall, England.
(Photograph provided by P. J. Pollard).

The spectacular broken onion-skin curved fragmentation is considered to be an extension of the curved sheeted fracture style represented in ***Plate 31***. The original curving fracture has been intruded by finely milled, dusty tourmaline-altered materials. Curved to rectangular fragments have been levered off into the main breccia flow and further comminuted and/or altered.

PLATE 31 Special features associated with selected intrusive breccias in curved sheeted fractures and alleged decompressive shock textures
(Photograph provided by P. J. Pollard).

PLATE 32 Special features associated with selected intrusive breccias – curved sheeted fractures and alleged decompressive shock textures
(Photograph provided by P. J. Pollard).

◻ Plate 33

Special features associated with selected intrusive breccias – sheeted fractures and shingle texture.
Wheal Remfry china-clay mine, Cornwall. England.
(Photograph provided by P. J. Pollard).

This feature of intrusive breccias is common within tourmaline related styles and is generally referred to as shingle texture. The shingle designation relates to the semi-parallel, overlapping, elongate pieces of wall rock, which appear to be spalling off from the porphyritic wall rocks. The elongate fragments are less distinct to the right and are both smaller and more altered to tourmaline (dark). At the extreme right the texture is changing to slightly more rounded fragmentation.

The origin of the shingle/curving texture is uncertain, but seems to relate to curved parallel sheet fractures (see *Plates 31 and 32*). Shingle texture is widely reported in tourmaline style intrusive brecciation, where explosive and rapid decompressive events are favoured interpretations.

◻ Plate 34

Special features associated with selected intrusive breccias – shingle texture.
Wheal Remfry china-clay mine, Cornwall, England.
(Photograph provided by P. J. Pollard).

Another example of shingle style brecciation, with interpreted disruption via a high-pressure volatile intrusive tourmaline-altered rock flour matrix. Note the degree of fragment rounding is very minor and there is a considerable amount of material which can be visualized as a jig-saw fit (limited transport, but high dilation). See also *Plates 13, 31, 32, 33* for sheeted fracture effects.

PLATE 33 Special features associated with selected intrusive breccias – sheeted fractures and shingle textures. Wheal Remfry china-clay mine, Cornwall, England (Photograph provided by P. J. Pollard).

PLATE 34 Special features associated with selected intrusive breccias – shingle texture. Wheal Remfry china-clay mine, Cornwall, England (Photograph provided by P. J. Pollard).

Plate 35

Special features associated with selected intrusive breccia – limited matrix. Welcome gold mine, Mingela, Queensland, Australia.

This highly photogenic specimen is from a small 40 × 30 m scale irregularly shaped pipe zone, which is interpreted to represent an unusual form of intrusive breccia.

The breccia is unusual in being predominately composed of very rounded fragments and large amounts of infill, but little matrix.

The fragments are of variably altered porphyritic granodiorite. The alteration affects the matrix of the igneous rocks, and where more advanced turns the rock grey-white with the phenocryst margins becoming blurred and difficult to discern. The alteration is of sericite ± silica style.

The prominent red hematite appears to be a different later-stage alteration effect (weathering)?

The prominent infill is composed of early crystals of sphalerite (dark) succeeded by irregularly distributed pyrite-marcasite (dull yellow) and chalcopyrite (yellow), with late-stage carbonate (white and grey). Gold is presumed to accompany this assemblage.

The remaining component is represented by the fine-grained, white-grey silicified material occurring sporadically between fragments. In places this exhibits poorly formed layering and/or more sandy domains. It appears to be a silicified matrix and poses some problems as

1. It seems too fine-grained to represent a normal milled matrix
2. It is very rare within the main outcrop.

One suggestion envisages the matrix as a tuffisite style remnant, with the bulk of the original milled matrix having been "blown away"? (See also tuffisite style, ***Plate 21 ABC***).

The timing of infill development is also enigmatic, and it is possible that it substantively postdates the fragment/matrix components and that very high fluid pressures have fractured and dilated rocks such that minimal matrix is now visible? The fragment, "matrix", infill relationships are exhibited again in the lower inset.

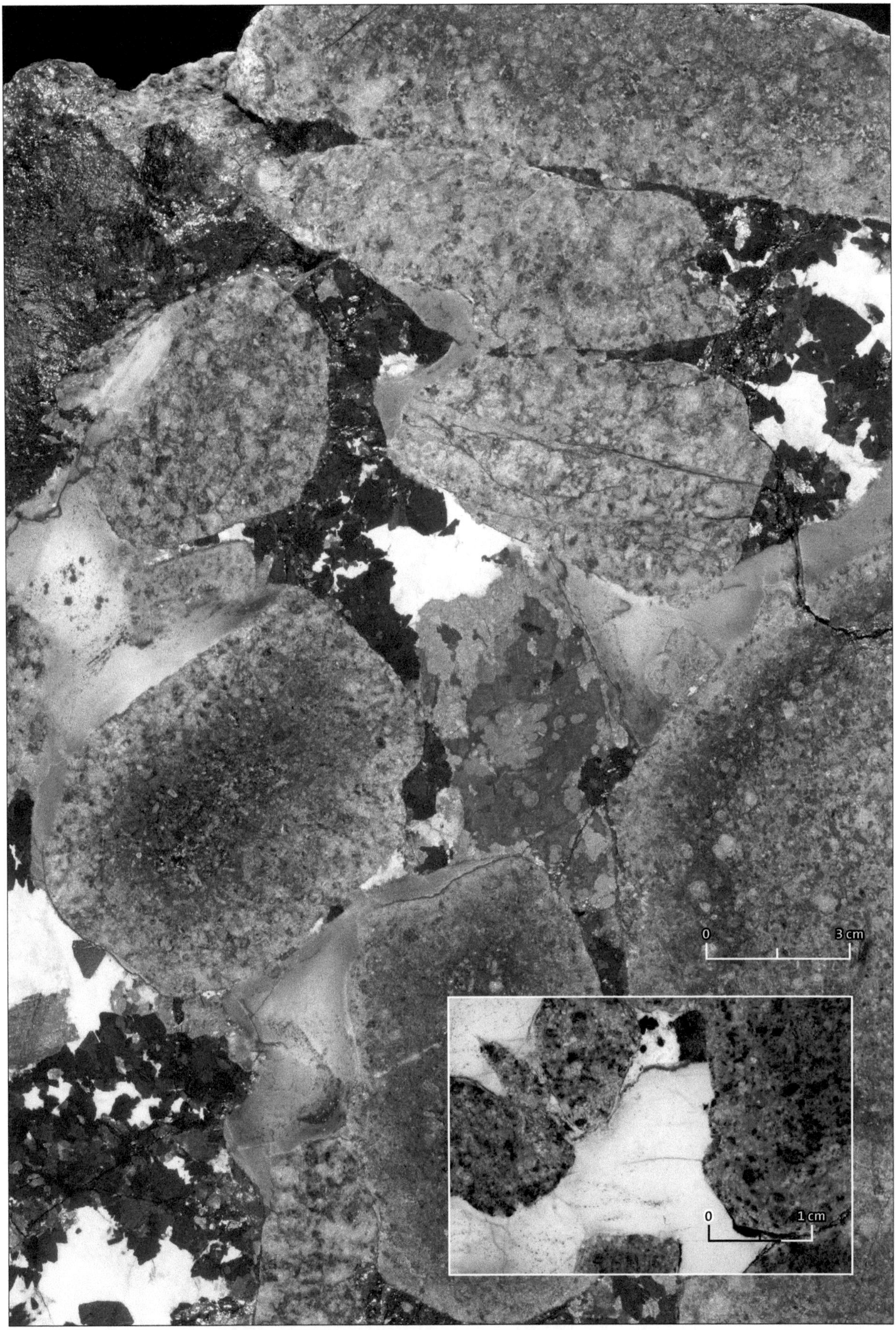

PLATE 35 Special features associated with selected intrusive breccia – limited matrix. Welcome gold mine, Mingela, Queensland, Australia.

◘ Plate 36 A B C

Special features associated with selected intrusive breccias – association with collapse breccia.
Ardlethan tine mine, New south Wales, Australia (White Crystal pit).
(Photographs provided by G. W. Clarke).

A The exposure represents the side wall of a major breccia body that has been mined for tin. The upper pit walls (paler colours) are composed of slightly altered (argillised) granite which passes downwards into the dark (tourmalinised) cassiterite-bearing rocks. The original central zone of the pit was completely tourmalinised, however within the marginal zone, it is possible to pick out many large-scale fragments. These are the pale tabular to lensoid zones within the darker tourmalinised rocks.

B This plate is a closer view of one of the large slab like fragments shown in ***Plate 36A*** (note the hammer scale to the right). The exposure illustrates the numerous angular fragments. The orientation of the slab-style fragments has been measured, and they form an inward funnel pattern pointing to the centre of the pit. The area beneath the big slab consists of many smaller angular fragments which are partially tourmalinised (black) and separated by infill zones of tourmaline and quartz (white). Note that these veins can also be seen cutting the main slab. Drilling beneath the pit revealed a breccia of intrusive style, and it is concluded that this pit exposure represents a collapse zone above or adjacent to an intrusive breccia (pressure relaxation)? The resultant high permeability zone has channeled and focused later hydrothermal fluid flow. (*Clarke et al, 1985*).

C The small plate is of a hand specimen derived from the heavily tourmalinised zone. The granitic wall rocks are variably tourmalinised – silicified, and angular fragments are "separated" by large patches of white infilling quartz crystals. Cassiterite (not visible) occurs coating quartz crystals.

PLATE 36 C

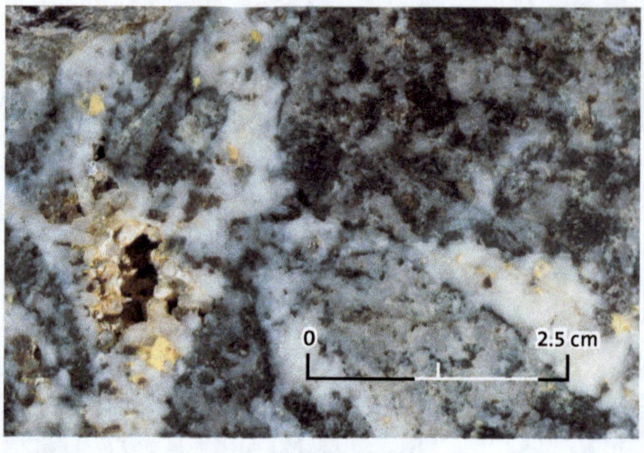

PLATE 36 A

PLATE 36 A B C
Special features associated with selected intrusive breccias – association with collapse breccia. Ardlethan tine mine, New south Wales, Australia (White Crystal pit)
(Photographs provided by G. W. Clarke).

PLATE 36 B

5 Intrusive Breccia Associated with Maar-Volcanoes and Associated Diatreme Structures

The term diatreme was originally utilised to describe a vent produced by volcanic explosion (Daubrée, 1891) and in recent times several epithermal systems have been identified associated with diatremes in a maar-volcano setting (*Sillitoe et al, 1984, Sillitoe, 1985*). The maar-volcano setting consists of a kilometre-scale circular crater-pipe structure surrounded by a ring of pyroclastic base surge and fall deposits. The system is considered to form via repeated phreatomagmatic explosions, with interaction between magma and groundwater creating the explosive energy.

The resultant crater is extremely unstable with slumping and faulting via ring faults partially infilling the pipe, which may ultimately or temporarily become the site of a crater lake.

The crater-pipes extend downwards to at least a kilometre and at lower levels consist largely of milled fragments with major matrix development (comminuted rock flow). In the upper sectors the pipe may consist of combinations of intrusive breccia, pyroclastic debris, possible tuffaceous (crystal) components, slumped and/or faulted blocks of wall rocks, organic debris, and occasionally glassy magma fragments. The wall rock components may include portions of the tuff ring with characteristic base surge style materials, accretionary lapilli and even lacustrine sediment. In general terms the combination of these features allows the intrusive breccia component to be placed in a diatreme context, and this section illustrates some of the textural features of observational value.

As a note of caution, the phreomagmatic origin should not be an automatic assumption with some breccia experts (including the author) suspecting that near surface magmatic fluid release might also cause a similar result (see also *Sillitoe, 1985*). It should also be noted that there are many examples of diatreme styles that are not associated with mineralisation.

◻ Plate 37

Diatreme domains – general surficial appearance, lacustrine sediments. Grasberg (copper, gold) mine, Papua, Indonesia.

This late-stage fragmental body occurs as a crescent shaped zone in the southwest region of the open pit and is devoid of copper-gold mineralisation. The general surface dimensions are some 8 × 70 m and it extends downward for at least 150 m. Smaller drill intercepts within the surrounding carbonate-dominated metasediments suggest dyke-like bodies and the unit has recently been interpreted as part of a late diatreme style unit *(Prendergast 2001)*.

Factors favouring this interpretation are evidence of major milling with a high matrix component, the general shape and the presence of argillised finely laminated contorted slabs of sediment (lower centre). The latter are considered to be slumped blocks of lake sediment. The exposure is extremely soft/muddy and unconsolidated, with both supergene and hypogene argillisation suspected. Within the plate the breccia body is unconformably overlain by well-layered sediments including a basal conglomerate.

◻ Plate 38

Diatreme domains – general surface appearance"muddy". Grasberg (copper, gold) mine, Papua, Indonesia.

An enlargement of the finer grained fragment-matrix component present within ***Plate 37***. This illustration is photographed some 20 m along strike from the above location and is more typical of the overall texture.

Centimetre-scale fragments of carbonate and argillised metasediment are supported by a major fine-grained matrix component. Evidence of fragment mixing, comminution and rounding is abundant. The exposure is very muddy, soft and essentially unconsolidated.

PLATE 37 Diatreme domains – general surficial appearance, lacustrine sediments. Grasberg (copper, gold) mine, Irian Jaya, Indonesia.

PLATE 38 Diatreme domains – general surface appearance "muddy" Grasberg (copper, gold) mine, Irian Jaya, Indonesia.

Plate 39

Diatreme domains – common style of intrusive breccia in diatreme environments. Wau gold field, Papua New Guinea. (Photograph provided by E. M. Baker).

The epithermal gold mineralisation of the Wau district is considered to be associated with the generation of a maar volcano and associated endogenous domes (*Sillitoe et al., 1984*).

The plate illustrates a mix of angular and rounded clasts set within a fine grained matrix. The dominant fine matrix, fragment rounding and general geological setting, have enabled the interpretation that this intrusive breccia is of diatreme style (see also ***Plates 40, 41***).

Plate 40

Diatreme domains – common styles of fragmental rocks, organic layering. Wau gold field, Papua New Guinea. (Photograph provided by E. M. Baker).

This fragmental rock has been interpreted (*Sillitoe et al., 1984*) to represent part of the upper portion of a maar vent/diatreme. The interpretation is based upon the geological context of the region. In this case suspicions are aroused by the presence of a block of interbedded finely layered siltstone with dark organic layers contained within breccia consisting of angular to rounded fragments with significant finer grained matrix components. The organic rich siltstone could be a product of slumping or fall into the diatreme vent. The exposure is very near to surface and both soft and argillised. (See also ***Plate 38***).

PLATE 39 Diatreme domains – common style of intrusive breccia in diatreme environments.
Wau gold field, Papua New Guinea
(Photograph provided by E. M. Baker).

Diatreme domains – common styles of fragmental rocks – organic layering.
Wau gold field, Papua New Guinea.
PLATE 40 (Photograph provided by E. M. Baker).

◻ Plate 41 A B C D E

Diatreme domains – general textural aspects of value in interpretation of diatreme related intrusive brecciation.
A, B, D – (Photograph provided by. E. M. Baker).
E – (Photograph provided by. G. W. Morrison).

- A Accretionary lapilli, Lepanto gold district, Philippines. These accretionary lapilli are of the fine grained argillised dust style and consist of vaguely concentrically banded fine grained ash within a tuff bed. Given that they are thought to form via accretion of dust onto water droplets in expanding phreatomagmatic explosive clouds, some form of diatreme system is indicated within the local district. This is indeed true and the deposits also have a spatial and temporal association with epithermal gold on the regional scale.
- B Accretionary lapilli, Wau gold district, Papua New Guinea. Accretionary lapilli within tuffaceous fragmental rocks interpreted to represent pyroclastic "tuff-ring" rocks of a maar-volcano/diatreme system (*Sillitoe et al, 1984*). Note the various styles of concentric layering in the pellet size particles.
- C Accretionary lapilli, Kelian gold mine, Indonesia. Accretionary lapilli concentrated within a dark layer of tuffaceous rocks. These are very difficult to see in field situations and are only just apparent here subsequent to slabbing and grinding. A selection of lapilli are outlined in the lower representation of ***Plate 41C*** to guide visual recognition. The unit suggests the nearby presence of a diatreme style vent (the Kelian breccia pipe).
- D Wood fragment, Hornsby diatreme, New South Wales, Australia. This specimen was collected from the vent/pipe zone of an unmineralised diatreme. It illustrates the nature of the matrix dominated intrusive breccia, with tiny scale pellet-like fragments set within a major amount of very fine ("muddy") comminuted clay/grit size matrix. This material upon reaching surface would have deposited as pyroclastic base surge and fall materials, and to the average geologist would merit the terminology of tuff. This again illustrates the problems involved in identifying intrusive breccia. The prominent banana-shaped fragment is actually a piece of carbonised wood, enabling a "near surface" interpretation.
- E Base surge deposition, Kilbourne Hole, United States of America. Typical base surge material of pyroclastic materials. Note low angle cross bedding of less the 15°, repetitive nature, and graded bedding. This style of material is common with the "tuff ring" of maar-diatreme volcanoes, and in combination with other features assists in establishing the diatreme intrusive breccia environment.

PLATE 41 A B C D E
Diatreme domains – general textural aspects of value in interpretation of diatreme related intrusive brecciation.
A, B, D – (Photograph provided by E. M. Baker).
E – (Photograph provided by G. W. Morrison).

PLATE 41 A

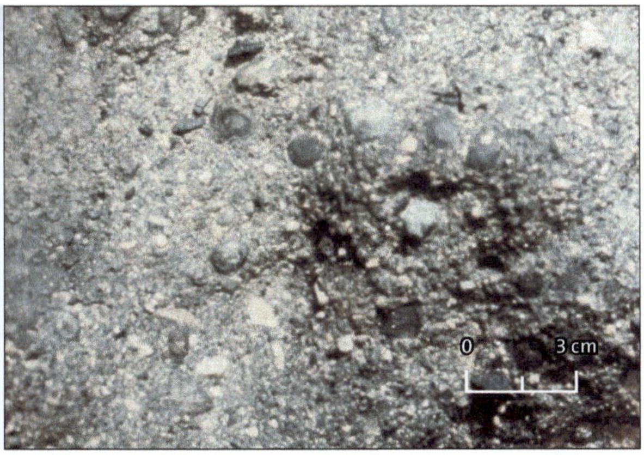

PLATE 41 B

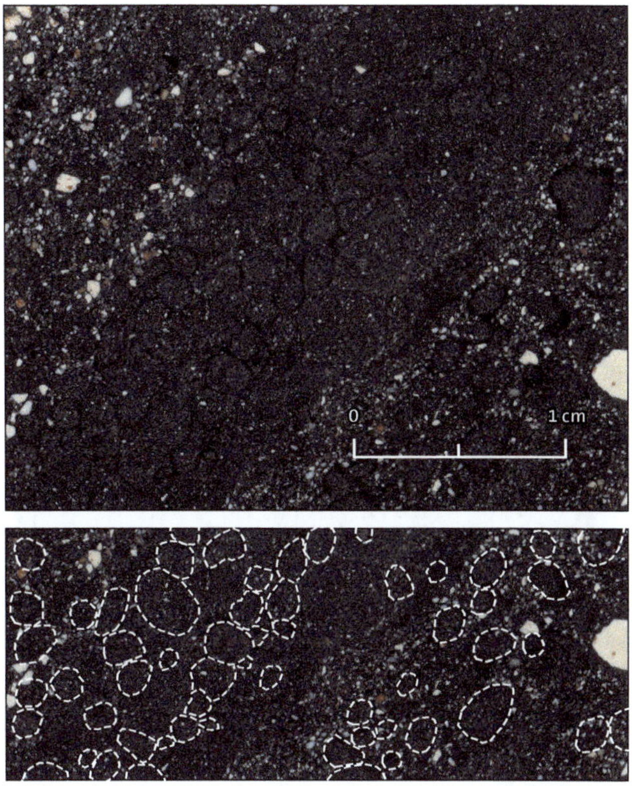

PLATE 41 C

PLATE 41 D

PLATE 41 E

6 Intrusive Breccia in Other Epithermal Systems

This volume has concentrated upon porphyry related intrusive breccia styles, and this small section is included as a reminder that many styles of intrusive breccia are present within epithermal domains. The maar volcano setting is discussed on ***page 274***, and two more smaller scale examples are illustrated here. The phenomenon of within-vein intrusive breccia is far more common than generally appreciated, and is possibly related to sudden fluid expansion/degassing upon localized pressure changes?

Plate 42

Intrusive breccia in epithermal settings – quartz vein. Woolgar gold field, Queensland, Australia.

The pre-existing epithermal quartz vein has been extensively shattered, creating many angular rotated fragments. However close inspection reveals a considerable proportion of rounded fragments at the medium to small scales. This suggests a milling/intrusive breccia component. The finer scale-matrix fragments are partially replaced by darker silica, and it is probable that even smaller scale grit/dust matrix particles have been completely replaced.

The original quartz vein contains some small scale and slightly obscure examples of silicified bladed carbonate textures which are characteristic of many low sulphidation epithermal systems (within the large fragment – arrowed).

Plate 43 AB

Intrusive breccia in epithermal settings – small scale intrusive breccia, silicification. Wirralie gold mine, Queensland, Australia.

Finely milled tuffaceous volcanic material (pale) has intruded and dilated (***Plate 42A***) the fractured tuffaceous rocks. Subsequent or associated dark siliceous alteration has obscured the texture of the original matrix components. A few (dark) siliceous fragments are still visible (***Plate 42B***) and represent an earlier quartz vein event. Late stage quartz veining is also represented (***Plate 42B***) with dark grey quartz veinlets overprinting the breccia. Matrix replacement of this extensive nature is a common feature of epithermal related intrusive brecciation. Significant gold values are commonly associated with dark silica stages within epithermal systems.

PLATE 43 B

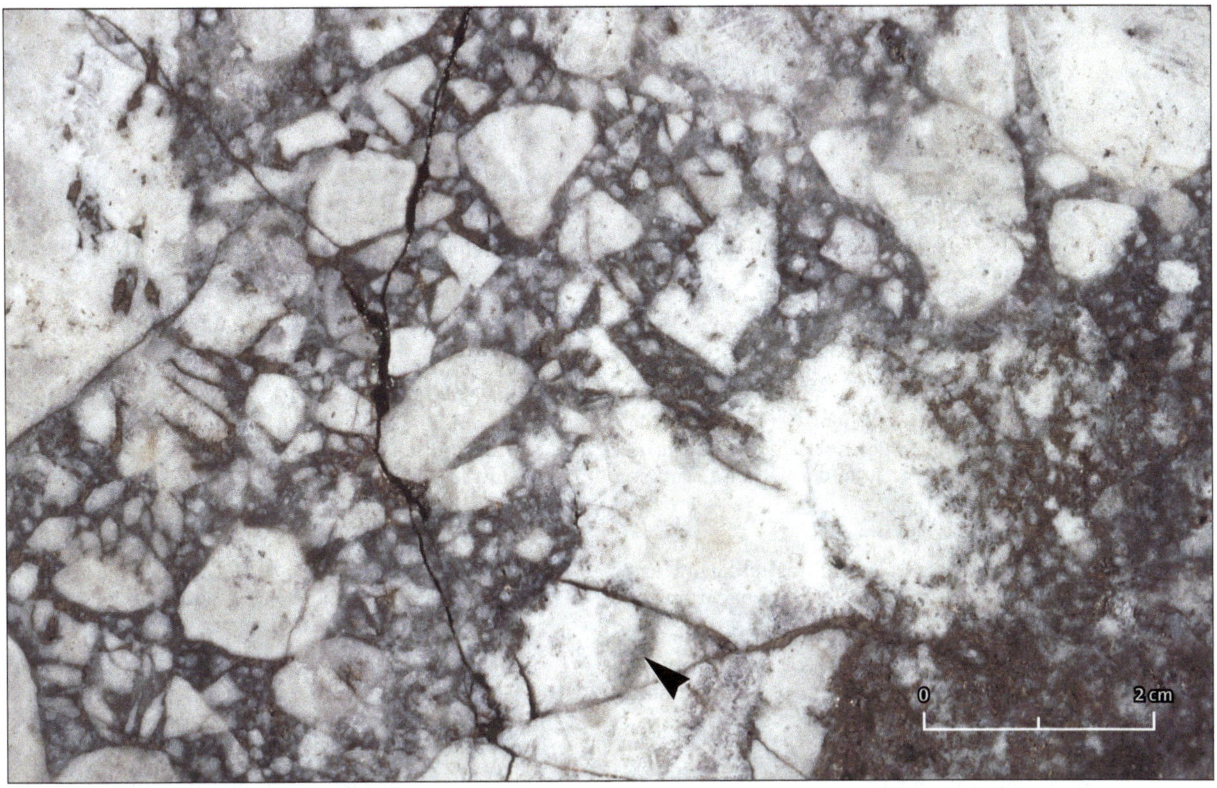

PLATE 42 Intrusive breccia in epithermal settings – quartz vein.
Woolgar gold field, Queensland, Australia.

Intrusive breccia in epithermal settings – small scale intrusive breccia silicification.
PLATE 43A Wirralie gold mine, Queensland, Australia.

7 References

Allman-Ward, P., Halls, C., Rankin A., and Bristow, C. M., 1982. An intrusive hydrothermal breccia body at Wheal Remfry in the western part of the St Austell granite pluton, Cornwall, England. *In* Evans, A. M. (ed.), Metallization associated with acid magmatism. Chichester, John Wiley and Sons, p. 1–28.

Baker, E. M., and Andrew, A. S., 1991, Geologic, fluid inclusion and stable isotope studies of the gold-bearing breccia pipe at Kidston, Queensland, Australia. Economic Geology, v. 86, p. 810–830.

Broadhurst, E., 1953. The Chillagoe copper lead field. *In* Edwards, A. B. (ed.), Geology of Australian ore deposits. Melbourne, Australasian Institute of Mining and Metallurgy, p. 768–782.

Bryant, D. G., 1968. Intrusive breccias associated with ore, Warren (Bisbee) mining district, Arizona. Economic Geology, v. 63, p. 1–12.

Burnham, C W., 1985. Energy release in subvolcanic environments: Implications for breccia formation. Economic Geology, v. 80, p. 1515–1522.

Clarke, G. W., Paterson, R. G., and Taylor, R. G., 1985. The nature and origin of brecciation and mineralisation at the White Crystal ore deposit, Ardlethan tin mine, New South Wales. Australian Journal of Earth Sciences, v. 32, p. 343–348.

Daubree, A., 1891. Recherches experimentales sur le rôle des gaz e hautes temperatures. Bulletin de la Société Géologique de France, v. 19, p. 313–354.

Hedenquist, J. W., and Henley, R. W., 1985. Hydrothermal eruptions in the Waiotapu geothermal system, New Zealand. Their origin, associated breccias and relation to previous metal mineralisation. Economic Geology, v. 80, p. 1640–1668.

Prendergast, K., 2001. Evolution of the West Grasberg breccia zone and gold bearing limestone breccias, Ertzberg Mining district. *In* Williams, P. J. (ed.), A hydrothermal Odyssey: Extended conference Abstracts, Economic Geology Research Unit, Contribution 59, School of Earth Sciences, James Cook University, p. 168–169.

Sheridan, M. F., and Wohletz, K. H., 1981. Hydrovolcanisms: basic considerations and review. Journal of Volcanology and Geothermal Research, v. 17, p. 1–29.

Sillitoe, R. H., Baker, E. M., and Brook, W. A., 1984. Gold deposits and hydrothermal eruption breccias associated with a maar volcano at Wau, Papua New Guinea. Economic Geology, v. 79, p. 638–655.

Sillitoe, R. H., 1985. Ore related breccias in volcanoplutonic arcs. Economic Geology, v. 80, p. 1467–1814.

Skewes, M. A., Arevalo, A., Floody, R., Zuniga, P. H., and Stern, C R., 2002. The giant EI Teniente breccia deposit – hypogene copper distribution and emplacement. Society of Economic Geologists, Special Publication 9, p. 299–332.

Ore Textures – Index

Index Front Page

Specimen from the Ravenswood gold mining district, Queensland, Australia (SYC Pit? Carpentaria Gold). This specimen illustrates the common situation of repeated re-opening and infilling of the same fracture system. It also illustrates just how difficult this can be to decipher. The problem is compounded when each stage contains similar components (see also page 147).

Index

F

G

H

I

J

K

L

M

T

U

V

W

Z

MIX
Papier aus verantwortungsvollen Quellen
Paper from responsible sources
FSC® C105338

If you have any concerns about our products,
you can contact us on
ProductSafety@springernature.com

In case Publisher is established outside the EU,
the EU authorized representative is:
Springer Nature Customer Service Center GmbH
Europaplatz 3, 69115 Heidelberg, Germany

Printed by Libri Plureos GmbH
in Hamburg, Germany